Framework

NCE

ldy Gannon

OXFORD
UNIVERSITY PRESS

OXFORD
UNIVERSITY PRESS

Great Clarendon Street, Oxford OX2 6DP

Oxford University Press is a department of the University of Oxford.
It furthers the University's objective of excellence in research, scholarship,
and education by publishing worldwide in

Oxford   New York

Auckland   Cape Town   Dar es Salaam   Hong Kong   Karachi
Kuala Lumpur   Madrid   Melbourne   Mexico City   Nairobi
New Delhi   Shanghai   Taipei   Toronto

With offices in

Argentina   Austria   Brazil   Chile   Czech Republic   France   Greece
Guatemala   Hungary   Italy   Japan   Poland   Portugal   Singapore
South Korea   Switzerland   Thailand   Turkey   Ukraine   Vietnam

Oxford is a registered trade mark of Oxford University Press
in the UK and in certain other countries

© Paddy Gannon 2004

The moral rights of the author have been asserted

Database right Oxford University Press (maker)

First published 2004

All rights reserved. No part of this publication may be reproduced,
stored in a retrieval system, or transmitted, in any form or by any means,
without the prior permission in writing of Oxford University Press,
or as expressly permitted by law, or under terms agreed with the appropriate
reprographics rights organization. Enquiries concerning reproduction
outside the scope of the above should be sent to the Rights Department,
Oxford University Press, at the address above

You must not circulate this book in any other binding or cover
and you must impose this same condition on any acquirer

British Library Cataloguing in Publication Data

Data available

ISBN 13: 978-0-19-914899-6
ISBN 10: 0-19-914899-8

10 9 8 7 6 5 4

Printed in Italy by Rotolito Lombarda

Author dedication
In loving memory of Pat Gannon

Contents

Introduction

Science is the study of the material and workings of our universe. It helps you to understand things that happen, and also to predict what might happen in the future. Looking at scientific evidence allows you to form views about changes in this technological world of ours.

But there's an even better reason to study science – it's good fun! This book aims to help your learning in science as part of a course including lots of practical work and activities. We hope that you enjoy it.

The dreaded SATs

In Year 9 you take your SATs, tests which measure your ability against students in the rest of the country. You will be tested on the science that you have covered in Years 7, 8 and 9, so there is a fair bit to learn. Make sure you understand all the new ideas as you come across them. You can use this book along with the Year 7 and 8 books to help you revise.

How to use this book

If you want to find out about a particular bit of science use the Contents and Index. There's also a **Glossary** (page 151) which gives you the meanings of most of the science words you'll meet in your Year 9 course.

The book is divided into 12 topics. At the start of each one is an **opener page** reminding you of what you already know about a topic, and a summary of the key ideas to come.

Then there is a set of double-page spreads, with questions for you to try and a **Language bank** of important words. The spreads are labelled **A**, **M** or **S**. **A** spreads introduce the easier ideas, **M** spreads follow on from these, and **S** spreads include some things which may make you think a bit harder.

At the end of the topic there are **Checkpoints** with questions to test yourself with, so you can find out if there are any ideas you need another look at before moving on.

But if you are still unsure about something, then ask your teacher to explain it again – they don't bite, honest!

Inheritance and selection

Before starting this unit, you should already be familiar with these ideas from earlier work.

- Organisms that are very similar are grouped as a **species**. But individuals in the same species are still different from each other.
- Some of the **characteristics** that make us all different came from our parents. Others are affected by our surroundings.
- In **sexual reproduction**, a male and female cell fuse together. What do we call this fusing together?
- Sex cells are **specialised**. In humans, the egg is specialised for storing food and growing into a baby. What is the sperm specialised for?

You will meet these key ideas as you work through this unit. Have a quick look now, and at the end of the unit read them through slowly.

- There are **genes** in the nucleus of every cell, and these control the characteristics of the organism.
- During fertilisation, genes from one parent join with genes from the other parent. The genes mix and the new individual has its own new set of genes.
- Some characteristics are **inherited** – they are passed on in the genes from parent to offspring. An example is eye colour.
- Other characteristics are affected by the **environment** or things that happen to you. For example, pierced ears are not an inherited characteristic.
- Some features are affected by both genes and the environment. You may inherit genes for being intelligent, but if you are not educated you will not do so well in an intelligence test.
- Some organisms have characteristics that make them successful. For example, a very fierce, strong and fast lion will catch lots of food. It is likely to thrive and pass its genes onto its cubs.
- Genes can be selected by people instead. People often choose organisms with useful characteristics and breed them so their genes will be passed on. This is called **selective breeding**.
- The characteristics that are useful to people depend on what we are trying to achieve. A dairy farmer selects cows that produce lots of milk. A beef farmer selects cows that produce good meat.

Inheritance

- What characteristics can be inherited?
- Why are offspring of the same parents similar but not identical?
- How do differences between offspring with the same parents compare with differences between offspring of different parents?

Genes from their parents have made sure these puppies are border collies not terriers. The puppies are all similar to their parents, but they are all slightly different.

Genetic and environmental variation

In the family photo, you can see that some of the people are related by blood (genetically) and some are related by marriage. The people who are genetically related show more similarities because they have inherited some similar characteristics. **Inherited variation** is differences between living things that are caused by **genes**.

The characteristics that are similar in family members are **inherited characteristics**, such as hair and eye colour and shape of nose. Brothers and sisters inherit similar features from their parents. But each person inherits a slightly different combination – brothers and sisters may all have 'family noses' but different colour hair. Everyone is unique.

Other differences between us have nothing to do with our genes. For example, characteristics such short or long hair, a scar on our left knee or a tattoo on our wrist are not inherited from our parents. These different features depend on our surroundings or on things that happen to us, and they are called **environmental variation**.

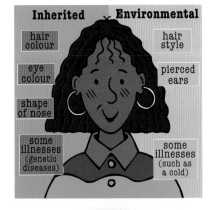

How do we inherit characteristics from our parents?

The **genes** carry the information about our inherited characteristics from one generation to the next. Genes are in the nuclei of our cells.

Your life started when your father's **sperm** fertilised your mother's **egg**. The nucleus of the sperm and the egg each carry their own set of genes. These genes are combined in the new person. The baby will have its own set of characteristics, mixed up from those of its parents.

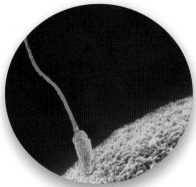

Fertilisation *produces a new life that is similar to its parents, but with its own mix of genes from the nucleus of the sperm and the nucleus of the egg.*

Plant features

Fertilisation also happens in plants. The male **pollen** fertilises the female **egg cell** inside the ovule. A seed develops, which contains its own combination of genes.

Measuring variation

We can measure the value of some features, such as people's height, mass, hand span or length of right foot. These features are usually affected by the environment.

For other features we can ask questions with just a few possible answers, such as:

○ 'What sex are you?' (Answer: male or female)

○ 'What blood group are you?' (Answer: A, B, AB or O).

These features are affected by genes – they are inherited features.

We can draw graphs to show these different types of variation.

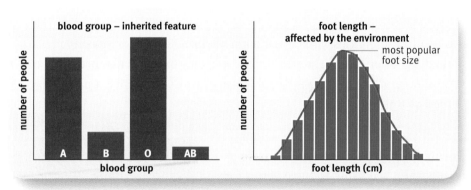

For many inherited features, there are just a few possible values and we draw a bar chart. For features that are affected by the environment, there are lots of values in between and we can draw a smooth curve.

Some characteristics, such as height, are affected by both genes and the environment. You may have genes for being tall, but if you don't eat a healthy diet you will not grow to be so tall.

1 Copy and complete using words from the Language bank:

Variation is differences in _____ between individuals. Characteristics passed to offspring from parents can result in _____. Characteristics that are affected by our surroundings lead to _____.

2 Look at the photos and describe some similarities that show inherited features in: **a** the family **b** the tomatoes.

3 List some human characteristics that are affected by the environment, and explain how they may be affected.

4 Jed thinks that plants don't show inherited variation. Draw labelled diagrams to explain fertilisation in plants, and why new plants are similar but not identical to their parents.

These tomatoes are different varieties. They have inherited genes for tomatoes with different characteristics.

Language bank

variation
characteristics
genes
genetic variation
inherited variation
environmental variation
offspring
sperm
egg
fertilisation
specialised
pollen

○ How are new breeds of animal produced?

Natural selection

Animals and plants show variation – they all have their own characteristics. This can make some individuals more successful than others. Imagine a male wolf that has genes for running very fast. He will be able to catch more prey to eat than other wolves. He is likely to survive, mate and have baby wolves, which may also be fast like their father. The characteristic of running fast has been selected. This is called **natural selection**.

Artificial selection

Sometimes people choose animals or plants that have useful characteristics and breed from them, to produce offspring with these same characteristics. This is called **artificial selection** or **selective breeding**. Over many generations people can use selective breeding to combine and exaggerate natural features until they have a particular **breed** of animal or plant.

For example, a farmer might want to produce a breed of sheep which give more meat. He would choose a ram with genes for lots of muscle, and mate it with muscly sheep. The offspring are likely to have genes for lots of muscle, and be useful to the farmer. If he continues this over many generations, he may end up with sheep that have quite a lot more muscle. He might also select woolly sheep to breed from, to end up with sheep that are woolly as well as meaty.

Guess what?

Farmers can buy ram semen from a supplier. It goes into the sheep's vagina through a tube, no intercourse needed! This is called artificial insemination.

Taming wild wolves

Do you fancy a pet wolf? Wolves are ferocious animals. Over millions of generations, people have used selective breeding to develop the useful characteristics of the wolf, and to remove less useful ones. This artificial selection has produced the many breeds of domestic dog that we know today. The table shows an example.

Over many generations people have selected parents and bred them together to produce the border collie, a breed useful for herding sheep.

Typical wolf characteristic	Border collie (sheep dog) characteristic
hates and fears people	friendly, loves people
hunts in packs and follows pack leader	follows human as pack leader
aggressive	good temperament
athletic, powerful, fast	athletic, good stamina, fast
intelligent, hunts by stealth	intelligent, obeys commands
has a drive to hunt prey	easily trained to herd sheep
frightening appearance	attractive appearance

Guess what?

*All breeds of dog are the same **species**. Only individuals from the same species can mate to produce fertile offspring. A horse and a donkey are different species. They can mate to make a mule, but a mule is not fertile so cannot reproduce.*

Breeders may have selected a fierce male wolf that was a good hunter, and mated it with a friendlier and more trusting female. They would hope to obtain some offspring that were less aggressive, but still athletic enough to work all day. By selecting these useful dogs and breeding from them over the generations, people developed a new breed of dog.

Selective breeding has produced many varieties of animal with particular features for specific jobs.

Gun dogs such as retrievers bring back the kill. Greyhounds were bred as hunting dogs, but people now bet on who will win the chase.

Waving wheat

People have also carried out selective breeding on plants for thousands of years. For example, wheat has been bred from wild grasses. Now the seeds heads are much bigger. Wheat stems are all the same height so that the ears come off easily during harvesting with machines.

Comparing natural and artificial selection

In natural selection, features are selected that help the individual survive, such as running fast. In artificial selection, features are selected that are useful to humans. This is not always best for the plant or animal – it does not always help it survive. Pug dogs have been bred to have a very short nose, and this can cause breathing problems. Sometimes genetic problems can get bred in along with the useful features – for example, dalmations are often deaf.

Modern wheat is the result of thousands of years of selective breeding.

1 Copy and complete using words from the Language bank:

In _____ breeding, people choose organisms with useful _____ to mate and pass on their _____ to the next generation.

2 Your mother is going to be a pig farmer. What characteristics might she want to select and breed for in her pig stock?

3 Describe how spots are passed from one generation of dalmation to the next, using the word 'cell' in your answer.

4 Some people say that selective breeding can weaken the species because it reduces the gene pool. Find out what this means and say whether you agree with the statement.

Language bank

artificial insemination
artificial selection
breed
characteristics
genes
natural selection
selective breeding
species
variation
variety

Breeding for a reason

Three hundred years ago, Hereford cattle were larger and carried a lot of fat. Their fatty meat suited people's tastes at that time.

○ Why do farmers produce new breeds of animal?
○ How are new varieties of plant produced?
○ Are varieties produced by selective breeding different from each other?

People breed animals for many purposes, including high milk or meat yields, fast growth, or being able to withstand bad weather. Plants may be bred for beautiful flowers, for flavour or for being able to resist pests and diseases. We develop new breeds mainly for economic reasons.

Leaner Hereford cows

The photos show how selective breeding has changed Hereford cattle, Over years of selective breeding Herefords became smaller.

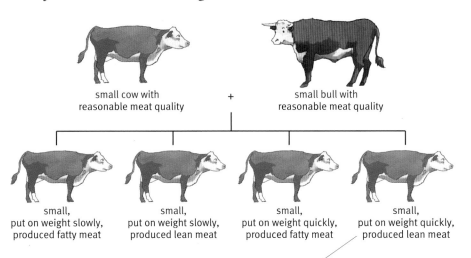

small cow with reasonable meat quality

+

small bull with reasonable meat quality

small, put on weight slowly, produced fatty meat

small, put on weight slowly, produced lean meat

small, put on weight quickly, produced fatty meat

small, put on weight quickly, produced lean meat

This animal was selected to breed from, and its offspring bred further to develop the characteristics of rapid weight gain and leaner meat.

Then people wanted meat with less fat. Breeders gradually developed leaner and trimmer beasts.

Plants with properties

The tables show some examples of tomatoes and lettuces with particular characteristics, produced by selective breeding.

You can grow tomatoes that are white, black, green, yellow, orange, pink, purple or striped!

plum tomato

beefsteak tomato

cherry tomatoes

Tomatoes	Characteristics
plum	Good for cooking or eating fresh. Medium sized with a fine thick flesh. Ideal for canning and reduce down well for sauces.
beefsteak	Good for slicing in sandwiches as they hold together well when cut. Big and juicy but flavour can vary.
cherry	Tiny fruit, used whole in salads, sweet. Rarely needs peeling, bright red colour.

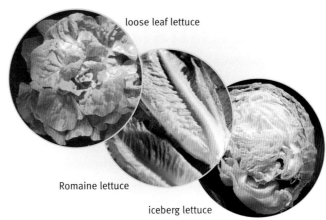

loose leaf lettuce

Lettuces	Characteristics
loose leaf	Cascade of loose leaves with no heart. Some with tender delicious flavour. Others have poorer taste and are used for decoration.
Romaine	Upright and columnar. Sweet creamy taste. White crispy heart surrounded by sturdy outer leaves. Hard to grow in poor soil; may not thrive in very hot weather.
iceberg	White crispy densely packed shape. Good heart taste. Grown in summer only.

Romaine lettuce

iceberg lettuce

Selective pollination

Remember pollination? Pollen travels from the stamen of a flower to a stigma. The pollen nucleus fertilises the egg.

In nature pollen is carried by the wind from one plant to another, or brought by insects. You can't tell which plant will pollinate another. How do plant breeders make sure the pollen and egg cells of their chosen plants will combine and breed together?

The answer is that breeders can pollinate their selected plants themselves. They take pollen from their selected 'male' plant and dust it onto the stigma of their 'female' plant, using a paintbrush or cotton bud. This is called **selective pollination**. They may cover the 'female' flower until they pollinate it, to make sure no other pollen can get in and pollinate it first.

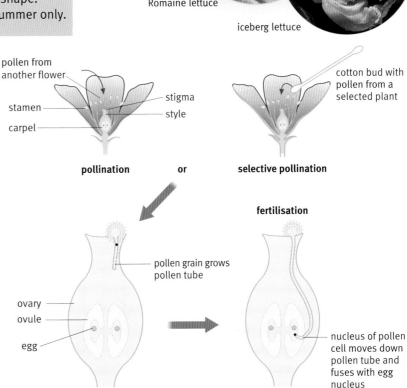

pollen from another flower

stamen

carpel

stigma

style

pollination or **selective pollination**

cotton bud with pollen from a selected plant

fertilisation

pollen grain grows pollen tube

ovary

ovule

egg

nucleus of pollen cell moves down pollen tube and fuses with egg nucleus

Selective pollination makes sure the chosen pollen fertilises the egg.

1 Copy and complete using words from the Language bank:

Farmers produce new _____ of animals and plants by the process of _____. They usually do this for _____ reasons.

2 List the different characteristics that you can see in the three types of lettuce.

3 List the characteristics farmers might select for in:
   a a milking cow   **b** a potato   **c** maize (sweetcorn).

4 Some tomatoes ripen later in the summer than others. What advantage might this be for a tomato grower?

Reproducing without sex

Asexual reproduction

In **sexual reproduction**, male and female sex cells combine and fertilisation happens. In a plant the pollen fertilises the egg in the ovule, and in an animal such as a human the sperm fertilises the egg.

Genes from both parents are combined in the offspring. Sexual reproduction produces variation – the offspring are all slightly different from each other and from their parents. In selective breeding, breeders try to minimise variation in the offspring, but even so they are not identical.

Sometimes plant growers want to produce offspring without variation. If they have a particularly good plant they may want to make lots more the same, without any chance of variation. Then they turn to **asexual reproduction**. This uses only one parent, and the offspring have genes the same as the parent's. Organisms with the same genes are called **clones**. Plant growers have been producing clones for many years, by several methods.

Do you think humans should be cloned?

Cuttings

To take a cutting, simply cut off a stem, remove a few leaves and stick it in soil. Roots soon develop at the end of the cutting and it grows into a new plant identical to the parent plant. Rooting powder helps the roots grow.

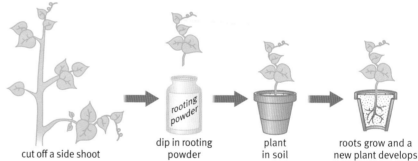

cut off a side shoot

rooting powder

dip in rooting powder

plant in soil

roots grow and a new plant develops

Growing mini-plants

Spider plants like this one on the left send out shoots and new plants grow at the ends of the shoots.

Grafting

Often an apple tree will produce nice juicy apples but it will not grow very strongly. To make lots of good apple trees, growers cut branches off the tree and join them to the stems of other trees that grow better. The grafted twigs grow and produce fruit, just like the original one.

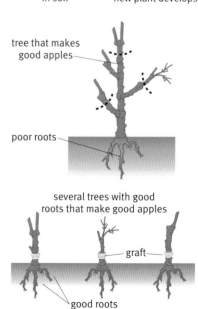

tree that makes good apples

poor roots

several trees with good roots that make good apples

graft

good roots

Pros and cons of cloning plants

Pros	Cons
identical offspring can be produced from one good stock plant	no variety, all have identical genes
offspring produced quickly without waiting for seeds to grow	a disease may wipe out a whole crop, not just a few weak individuals

Cloning animals

Cloning animals is not quite so simple.

Dolly the sheep is a clone of her mother who was produced like this:

o A nucleus was removed from one of the mother's body cells.
o The nucleus was removed from an unfertilised egg and the mother's body cell nucleus put into the egg.
o The egg developed into a sheep with genes identical to the mother.

Many people are very unhappy about the idea of cloning animals, particularly humans. In many countries it is illegal to try and clone humans. Imagine what would happen if one powerful person could fill the world with lots of people the same as them! There lots of other reasons why people are worried about human cloning.

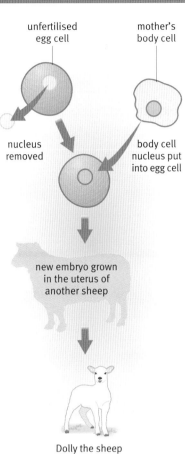

Dolly the sheep was cloned from her mother and was identical to her. However, she aged quickly and did not seem as healthy as her mother. Animal cloning is still at an early stage in its development.

1 Copy and complete using words from the Language bank:

In _____, male and female sex cells fuse. This is called fertilisation. The _____ of the parents combine and the offspring show _____. In _____, there is only one parent and the offspring are identical. They are called _____.

2 How can new plants be made without sexual reproduction?

3 Discuss the benefits and problems of cloning plants.

4 Find out some reasons why people are worried about human cloning.

Language bank

asexual reproduction
clones
cuttings
egg
fertilisation
genes
grafting
pollen
sexual reproduction
sperm
variation

Checkpoint

1 Plant fertilisation

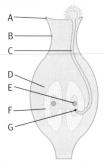

a Look at the diagram of fertilisation in plants. Match each label below to the correct letter.

pollen tube   stigma   style   ovule
pollen cell nucleus   egg   ovary

b Choose the correct reason why the new plants formed are similar but not identical to their parents.

- All the genes for the new plants come from the pollen grain.
- The genes for the new plants are a mixture from the pollen tube and the ovule.
- All the genes for the new plants come from the egg.
- The genes for the new plants are a mixture from the pollen nucleus and the egg nucleus.

2 All different

Copy and complete these sentences, unscrambling the words.

We are all slightly different – we show **naviatior**. We each have our own features or **tarcacterissich**. Some characteristics are **dienrieth**, which means they are passed on from our parents in the **seeng**. Others are affected by the **ronnivement** – our surroundings or what happens to us.

3 Inherited or environment?

Read this description of George. List his features, and decide whether you think each one is inherited, affected by the environment, or both.

George has red hair and blue eyes. He is quite tall and very skinny. He went on a football camp last year and is very good at football. His skin is pale but on holiday in Spain it turned pink. He has a straight nose and large ear lobes. One ear is pierced. He has a scar on his leg where he hurt himself when he was younger. He has shaved off one eyebrow and wants to put a blue streak in his hair.

4 Selective breeding

Match up each organism with the feature that a farmer might develop in selective breeding.

Organisms
chickens
beef cattle
wheat
milk cattle
daffodils
sheepdog

Features
strong stem of even length, big ears
good temperament, high milk yield
good temperament, fit and fast-moving
put on weight quickly, good muscle quality
strong stem, large yellow flowers
large eggs, long laying season

5 Selected by nature?

For the following features, decide whether they are more likely to be a result of natural selection or artificial selection (selective breeding).

- a short-nosed pug that has breathing difficulties
- a fast-running lion
- a very woolly sheep that doesn't wander away
- a long-necked giraffe
- a zebra with stripes
- a pig that produces lean meat

Fit and healthy

Before starting this unit, you should already be familiar with these ideas from earlier work.

- In order to carry out all the life processes, cells need energy. This comes from respiration, which is similar to combustion:

  glucose + oxygen → carbon dioxide + water   **Energy is given out.**

- When we burn a fuel, the energy is given out all at once. How is respiration in cells different from this?
- So cells can keep respiring, oxygen and carbon dioxide are exchanged in your lungs, inside tiny air sacs called alveoli.
- You need to eat a balanced diet to stay healthy and for your body to grow properly. List the seven components of a balanced diet.
- A developing fetus gets the oxygen and nutrients it needs from its mother's blood via the placenta.

You will meet these key ideas as you work through this unit. Have a quick look now, and at the end of the unit read them through slowly.

- For the body to be healthy, its organs and organ systems must work together. Cells need oxygen and glucose for respiration. The **respiratory system** provides oxygen and removes carbon dioxide, the **digestive system** provides nutrients and the **circulatory system** transports all these substances to and from cells.
- Smoking damages the respiratory and circulatory systems. Cancers caused by smoking can affect many organs, and smoking when pregnant harms the fetus.
- Deficiency diseases result when particular nutrients are missing from the diet. Obesity is a serious health problem caused by eating more food than the body needs.
- Alcohol is addictive, and drinking too much alcohol damages many organs including the liver and the brain. It can cause behavioural and social problems, and damage the unborn fetus.
- A **drug** is a substance that changes the way the mind or body works. Drugs have side effects and many are addictive, which can make them very dangerous. We classify drugs depending on how we use them. Nicotine, alcohol and illegal drugs are all harmful.
- Regular exercise improves fitness by making the respiratory and circulatory systems more efficient. Exercising too vigorously can sometimes damage the joints or muscles.

Fit for what?

○ What do we mean by fit?

Fighting fit

How fit are you? The answer is different for different people. Professional athletes and sports players work hard to be fit, but their ideas of fitness vary according to their sport. So what does it mean to be fit?

These people are all fit, but in different ways and for different purposes.

Remember respiration? To provide your body with energy for exercise, your cells respire. They need glucose and oxygen for this.

glucose + oxygen → carbon dioxide + water   **This releases energy**

Fitness is a measure of how well your body works. Your organs and organ systems work together so you can exercise:

○ Your **respiratory system** includes your lungs and airways. Here oxygen enters your body and carbon dioxide leaves (gas exchange).
○ Your **digestive system** includes your stomach and small intestine, providing your cells with digested food molecules such as glucose.
○ Your **circulatory system** includes your heart and blood vessels, which transport oxygen and glucose to all your cells, and carry wastes away.

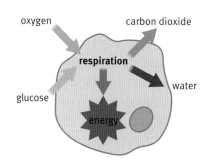

If you are fit, these organ systems work well together. You can exercise without becoming totally exhausted, and when you get out of breath you recover quickly. Being fit doesn't just help you with sport, it's helpful in everyday life as well.

To stay fit you need to look after your body. This means eating a balanced diet, and avoiding harming your body with alcohol, smoking or using drugs. Regular exercise helps keep you fit, making your heart and lungs bigger and more efficient, so they don't have to work so hard to provide your cells with oxygen.

how well your respiratory system works

how long your muscles can work for

how easily your body can move and bend

how much fat and muscle you have

There are many aspect of fitness. Different sports need different combinations of these.

How do we measure fitness?

If you are fit, your breathing rate and heart rate are low. During exercise they rise, but afterwards they return to normal very quickly. This is called the **recovery rate** and it is a good indicator of fitness.

To measure heart rate, you can feel the pulse in your wrist. The table shows what happened when three people did steps-ups for 3 minutes.

Patricia recovered quickly, because she is very fit. She exercises regularly and her body is good at providing her muscle cells with oxygen and glucose. Malcolm's body finds exercise very hard. But if he eats healthily and exercises regularly, his fitness level will soon improve.

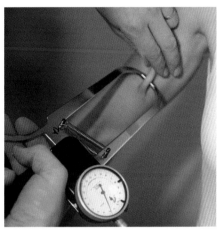

Too much fat makes exercise difficult. Skin fold callipers give an idea of how much fat you are carrying.

Guess what?

The Spanish cyclist Big Mig (Miguel Indurain) had a resting heart rate of 29 beats per minute and only 8% body fat. The average for a man is nearer 72 beats per minute and around 20% body fat.

Person	Pulse rate (beats per minute)	
	Before exercise	After exercise
Patricia	80	128
Leo	90	160
Malcolm	98	184

This device measures muscle strength. With training, muscles get bigger and stronger.

1 Copy and complete using words from the Language bank:

Fitness is your ability to exercise without becoming _____. It depends on eating a balanced _____, doing regular _____ and not abusing your body.

2 What is meant by recovery rate? Why is it a useful thing to measure?

3 **a** Write out the word equation for respiration.

b Explain how three organ systems work together so cells respire.

4 Mark likes eating chips, watching TV, drinking beer and smoking. He can't walk up a hill without feeling exhausted. Write a fitness programme for Mark, with notes on how it will help him.

Language bank

carbon dioxide
circulatory system
diet
digestive system
exercise
exhausted
fitness
glucose
recovery rate
respiration
respiratory system

Oxygen for respiration

○ What helps the respiratory system to function?

The **respiratory system** is the organ system that provides the oxygen you need for respiration, and removes the carbon dioxide that is formed. In the respiratory system, oxygen from the air goes into the blood. The circulatory system then transports the oxygen to all the body cells.

Breathing in and out

For gas exchange to continue, the air in the alveoli has to be continually replaced, to keep up the supply of oxygen and get rid of the poisonous carbon dioxide. This is the purpose of breathing, or **ventilation**.

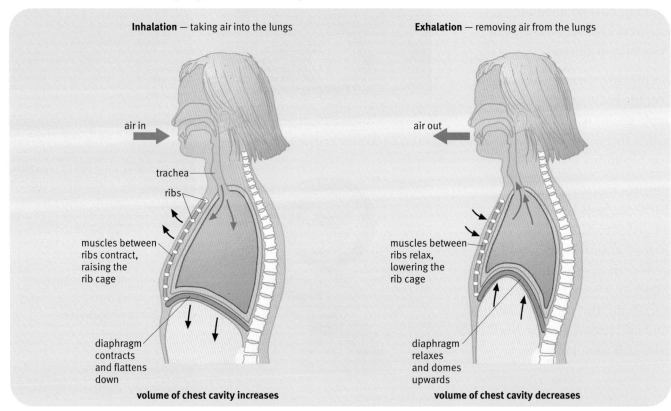

Inhalation — taking air into the lungs

air in

trachea
ribs
muscles between ribs contract, raising the rib cage
diaphragm contracts and flattens down

volume of chest cavity increases

Exhalation — removing air from the lungs

air out

muscles between ribs relax, lowering the rib cage
diaphragm relaxes and domes upwards

volume of chest cavity decreases

The lungs are in the **chest cavity**, enclosed in the **rib cage**. At the base of the chest cavity is a domed sheet of muscle called the **diaphragm**. When you breathe in (inhale), your ribs move up and out and your diaphragm flattens down. This makes the volume of the chest cavity increase, causing air to rush in through your nose or mouth, down your trachea and into your lungs.

When you breathe out (exhale), your ribs move down and in and your diaphragm relaxes and domes up again. This makes the volume of the chest cavity decrease, causing air to rush out of your lungs, up your trachea and out through your nose or mouth.

Do you breathe in pure oxygen and breathe out pure carbon dioxide?

No, you breathe air, which is a mixture. In inhaled air, 21% is oxygen and 0.04% is carbon dioxide. In exhaled air, only 16% is oxygen and 4% is carbon dioxide.

Demonstrating ventilation

  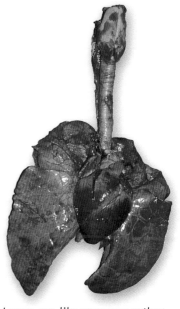

In this bell jar model of the lungs, pulling down the rubber sheet increases the volume inside the jar. Air rushes in through the pipe and the balloons inflate. Pushing up the rubber sheet decreases the volume inside the jar. Air rushes out of the balloons and they deflate.

Lungs are like sponges rather than balloons, but the bell jar demonstration helps you understand how they are ventilated.

Breathing and fitness

If you are fit, your lungs will have a large volume for your size. This means they can supply lots of oxygen to your blood. You also need a healthy circulatory system to carry the oxygen around your body.

Asthma is a condition in which the bronchioles that lead to the alveoli get narrower. This restricts the air getting into and out of the alveoli.

This peak flow meter measures how quickly you can force air out of your lungs. It can detect signs of respiratory problems such as asthma.

Guess what?

*Your **vital capacity** is the total amount of air you can breathe out after a deep breath in. The Spanish cyclist Miguel Indurain had a vital capacity of 7.1 dm^3, compared with the average for a man of 5.2 dm^3. A woman who is fit might have a vital capacity of 4.5 dm^3.*

1 Copy and complete using words from the Language bank:

In _____, air is taken into the lungs. The _____ flattens down and the rib muscles raise the ribs. This increases the volume of the _____ and air rushes into the lungs.

2 Write another passage like the one in question 1 for breathing out.

3 Explain how the bell jar experiment helps us to understand ventilation. Which part of the respiratory system does each part of the apparatus represent?

4 Find out more about asthma: what causes it? how is it treated? is it becoming more or less common?

Language bank

asthma
chest cavity
diaphragm
exhalation
gas exchange
inhalation
respiratory system
rib cage
ventilation
vital capacity

Smoking

○ What is the effect of smoking on the lungs and other body systems?

Your lungs are delicate organs that look like pink sponges. They have a very large area made up of tiny alveoli, so that you can exchange gases efficiently. These alveoli are easily damaged by air pollutants, especially those that come from smoking

In the smoker's lung on the right, the alveoli have been clogged up by smoking. How fit do you think this person would have felt?

What does smoking do for you?

There are no benefits from smoking. You might hear people say that it is cool or the 'in' thing to do, but the fact is that smoking is virtual suicide. It doesn't just make you unfit, it kills.

The table shows some of the poisons in cigarette smoke. The heat from the hot smoke also damages your airways and lungs.

Substance	What does it do?
carbon monoxide	A poisonous gas which prevents your red blood cells from carrying oxygen around the body.
nicotine	An **addictive** drug which makes it hard to stop smoking. It makes the heart beat faster and the blood vessels get narrower, causing high blood pressure. This can lead to heart disease and death.
tar	A black sticky substance which coats the surface of the alveoli, preventing gas exchange. Tar contains **carcinogens**, substances that cause cancer.

Pregnant women who smoke are more likely to have miscarriages, babies born dead or babies born with very low birth weights. Many of the poisonous substances in cigarette smoke pass through the placenta to the unborn baby.

Ciliated epithelial cells *line your airways and help keep them clean. The mucus traps any dust particles in inhaled air, and the cilia move the mucus and trapped dirt up away from the lungs. But tar in cigarette smoke destroys the cilia, so the mucus and dirt build up in the airways, leading to infections such as bronchitis.*

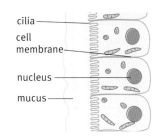

cilia
cell membrane
nucleus
mucus

This woman was healthy and attractive when she was young. She took up smoking. Here are some of her problems:

o Bronchitis (infected airways) and emphysema (her alveoli are breaking down). Constantly breathless, she now has to be on oxygen just to survive.

o Her leg had to be amputated because of a **thrombosis** (blood clot).

o She is at risk from blocked arteries. If the arteries supplying her heart are blocked, this will cause **angina** (pain in the chest) or even a **heart attack**, which could be fatal.

o She is at risk from **strokes**. A blood clot in the brain can lead to bleeding, damaging part of the brain and causing disability or death.

o She has yellow fingers and teeth and bad breath. Her skin aged earlier than normal.

o She has wasted thousands of pounds a year. Just think what else she could have spent all that money on.

Would *you* like her problems?

Passive smoking

Passive smoking means breathing in second-hand smoke from other people's cigarettes. This can cause eye irritation, headaches, coughs, sore throat, dizziness, nausea (feeling sick) and shortness of breath.

Why does this happen? In the smoke that you breathe in, there are 100 chemical poisons, including 50 known carcinogens.

Children whose parents smoke are more likely to suffer chest and ear infections than those with non-smoking parents. Non-smoking people who are exposed to passive smoking in pubs and clubs are at increased risk of cancer.

When she started smoking, the dangers were not so well known. Over the years people have collected more and more evidence, and today ignorance is not an excuse. We all know it's best not to start smoking.

Roy Castle never smoked but he died of lung cancer after working as an entertainer in clubs around the country. A charity in his name has been set up for lung cancer sufferers.

I don't understand why humans smoke, I'd be really sick if I tried it!

1 Copy and complete using words from the Language bank:

Smoking damages the _____. Smoke contains _____, which is a poisonous gas, nicotine, which is an _____ drug, and tar, which is a _____.

2 Gary and Chris are 15 years old and regularly meet behind the sports hall for a cigar, as they have heard that cigarettes are bad for you. Explain to them in detail where they are going wrong, and why any kind of smoking is dangerous.

3 Smoking can cause emphysema, testicular cancer and heart disease. Find out more about these problems. Explain why you would want to avoid them.

Language bank

addictive
cancer
carbon monoxide
carcinogen
ciliated epithelial cells
emphysema
nicotine
passive smoking
pollutants
respiratory system
tar

23

We are what we eat

○ Why is diet important?

To stay fit, you need a **balanced diet**. This provides all the **nutrients** in the correct proportions. Without the correct balance of nutrients your body does not have enough energy to carry out the life processes, and cannot grow and repair itself properly.

> **Remember**
> The seven types of nutrient are proteins, carbohydrates, fats, vitamins, minerals, fibre and water.

Deficiency diseases

If a nutrient is missing from someone's diet, such as a particular vitamin or mineral, they may suffer from a **deficiency disease**. The photos show some examples.

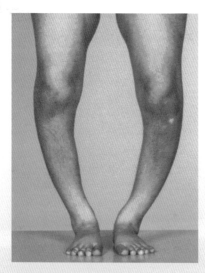

If vitamin D is deficient in childhood, rickets may result. The bones do not form properly. As well as eating vitamin D, we all need some sunlight so our bodies can make it.

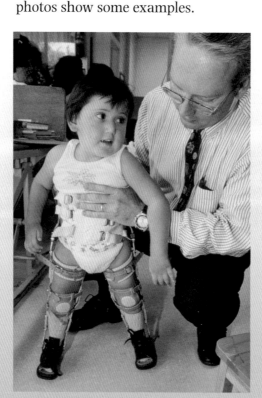

If a pregnant woman is deficient in a B vitamin called folic acid, the child may suffer spina bifida. The spine does not develop properly and this causes paralysis of the legs and other problems.

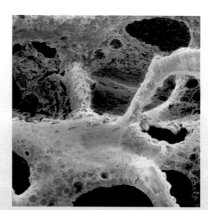

Lack of the mineral calcium can lead to osteoporosis. The bones become thin and break easily. The teeth, muscles and heart are also affected. This is very common in older people.

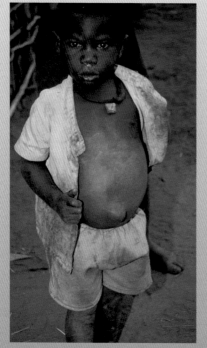

Without enough protein in the diet, a young child can suffer from malnutrition. Their body may swell but this is fluid, not fat. The child does not grow properly and is very unhealthy. Protein deficiency is common in parts of Africa, and it kills.

Lack of iron can cause anaemia. The blood cannot carry enough oxygen around the body. This makes people pale and they feel lethargic (very tired).

Obesity – the other side of the coin

In developed countries, malnutrition is not as common as it once was. However, many parents will know that getting children and teenagers to eat a balanced diet can be difficult. An increasing problem is **obesity**, another word for being overweight. Currently 15% of 15-year-olds are obese. Research suggests that if the 'fast-food couch-potato' culture of youth continues, the average life expectancy will start to fall. For hundreds of years until now, life expectancy has been rising – people are on average living longer than before.

Vital vitamins
Diseases such as scurvy and rickets have been known for hundreds of years, but learning what caused them was a challenge for doctors. Frederick Gowland Hopkins worked at Cambridge about 100 years ago, and was awarded the Nobel Prize for Medicine. His work linked the growth of animals to certain chemicals in their diet. He fed rats mixtures of different nutrients and soon realised they did not thrive simply on fat, carbohydrate, protein, minerals and water. They needed some other vital substances, now referred to as vitamins. It was the Polish-born Casimir Funk who first used the term 'vitamine' in 1912.

Some facts about vitamin C
- It is found in vegetables and especially citrus fruits like lemons and oranges.
- If is missing from the diet, it causes the deficiency disease scurvy.
- To test whether it is present in a food, we use DCPIP.
- It is water soluble and is destroyed by heat, so the longer you boil your vegetables, the less vitamin C will be left.

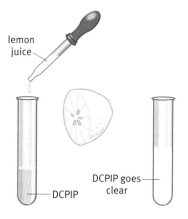

lemon juice

DCPIP goes clear

DCPIP

DCPIP is a blue dye which turns clear when vitamin C is added. We use this to test whether foods contain vitamin C.

1 Copy and complete using words from the Language bank:

To stay fit and healthy you need to eat a _____. This includes all seven _____ in the correct proportions. If a particular nutrient is missing, a _____ will result.

2 Make a table listing the seven kinds of nutrient. Add one column showing why we need each nutrient, and another column showing what kinds of food provide it.

3 Fast-food restaurants are sometimes criticised for targeting their advertising at children. Some people say these restaurants should be forced to include less salt, sugar and fat in their food, to make it healthier. What do you think about this?

4 Find out about a disease called anorexia: what is it? who suffers from it? why does it happen?

Language bank

anorexia
balanced diet
carbohydrate
deficiency disease
fat
mineral
nutrients
obesity
protein
roughage
vitamin
water

Alcohol

○ How does alcohol affect the body?

Alcohol is a **drug** – a substance that alters the way the body works. A drug might affect the mind, the body or both. Alcohol is called a **recreational drug** – people drink it to be sociable and to enjoy themselves. It is widely acceptable in Britain, although some cultures do not use it.

Alcohol changes the way people behave. It makes them feel relaxed, less shy and more talkative.

You may be surprised to learn that alcohol is a highly intoxicating and poisonous substance. When people drink it in moderate quantities, their bodies can deal with it by breaking it down and removing it before it causes serious damage or death.

What does alcohol do to the body?

Alcohol is an **addictive** drug. It can make people feel that they need to keep drinking. If they stop, they suffer **withdrawal symptoms** such as aches, pains, shaking and sweating. This makes them feel very ill until they have the next drink. Someone who is addicted to alcohol is called an **alcoholic**.

The **liver** deals with alcohol in the body, changing it into harmless substances that are removed from the body. If people drink heavily, the liver cannot cope and is damaged. Cirrhosis of the liver and cancer of the liver are both fatal diseases that are common in alcoholics.

Alcohol slows the nervous system. It is a **depressant**, which makes people feel relaxed. Too much alcohol can damage the brain, making it spongy.

Women who drink when they are pregnant can damage the developing fetus, particularly in the early stages of pregnancy.

Guess what?

The amount of alcohol people drink is measured in **units of alcohol**. One glass of wine or half a pint of beer or lager contains about 1 unit. Men are advised to drink no more than 21 units a week and women no more than 14 units, spread throughout the week.

Alcohol can make some people aggressive, and argumentative.

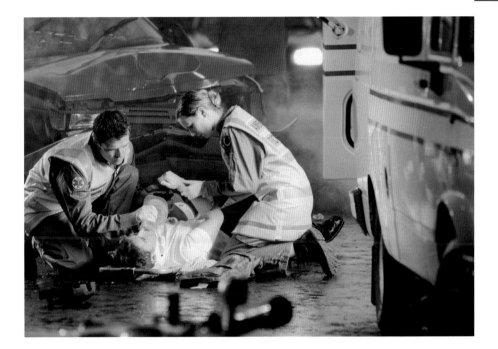

People who drive after drinking risk killing or injuring innocent people as well as themselves, The maximum prison sentence for causing death by drink-driving is 10 years.

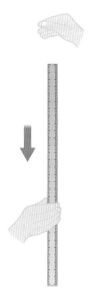

Social problems

By law, only those over 18 can buy alcohol. This is because like many benefits of being an adult, drinking alcohol carries a great responsibility. Doctors think that moderate sensible drinking may be healthy, helping prevent heart disease in people over 40. But for some people alcohol takes over their lives, gradually destroying them.

Alcohol makes your **reaction time** longer. This is the time between seeing a problem and your body reacting to it. Because of this, people who drive after drinking are much more likely to have an accident. Drunk drivers are responsible for some 600 deaths and thousands of serious injuries each year in the UK. Many families are torn apart by the loss of a relative killed by a drunk driver.

Young people who abuse alcohol run the risk of seriously damaging their health, or even death. Each year over 1000 people under the age of 15 are admitted to hospital with alcoholic poisoning, needing emergency treatment.

How far the ruler falls before it is caught indicates the reaction time. After just a few alcoholic drinks, the ruler falls further. Dropping a ruler is no problem; driving a vehicle with slowed reaction times can be deadly.

1 Copy and complete using words from the Language bank:

Alcohol is a _____ which changes people's behaviour. It can be _____, causing people to keep drinking too much. Alcohol affects people's _____, making it dangerous to drive.

2 Which two organs are damaged if people drink too much alcohol?

3 Explain what withdrawal symptoms are, and what causes them.

4 A friend has heard alcohol described as a social drug. They think this means that it is not dangerous. Write an email explaining how alcohol can be deadly.

Language bank

addictive
alcoholic
behaviour
brain
cancer
cirrhosis
depressant
drug
liver
reaction time
units of alcohol
withdrawal symptoms

Maintaining fitness

A healthy heart

Your heart powers the circulatory system, pumping blood carrying oxygen and nutrients around the body. Regular exercise makes the heart bigger and stronger, improving fitness.

There are several diseases of the circulatory system. Fatty deposits can form on the inside walls of arteries, which can lead to the arteries becoming blocked. If this happens in the brain, it can cause a **stroke**. If it happens in the arteries that supply the heart itself, it can lead to **angina** and a **heart attack**. The following can cause these fatty deposits to form:

- eating too much food containing animal fats, such as red meat, cheese, cream and butter
- drinking too much alcohol
- diabetes
- smoking
- genes – it can run in the family.

Another problem of the circulatory system is **high blood pressure**. This can result from deposits in the arteries, or from smoking, obesity, too much salt in the diet, or too little exercise. People with high blood pressure are more likely to have a stroke or heart attack, and may suffer kidney and eye problems.

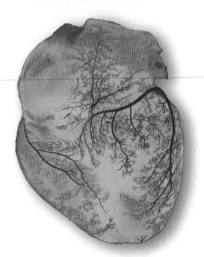

These arteries supply the heart with oxygen and glucose. If they become blocked the heart muscle can no longer respire, causing a heart attack which can be fatal.

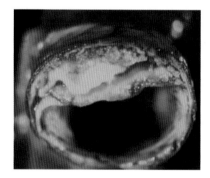

These white fatty deposits can build up inside arteries and block them.

Holding you up

Your **skeleton** supports your body, and allows you to move. It also protects delicate organs such as your brain and lungs. The diagram shows the bones in your skeleton.

Your bones are rigid, but you can move your body because you have **joints**. These are places where two bones meet. There are different types of joint, such as **ball-and-socket joints** at your hip and shoulder, and **hinge joints** at your knees, elbows and in your fingers.

In a joint, the bones are held together by strong fibres called **ligaments**. These make sure the bones move in the correct direction and do not move apart. On the surface of the bones is a protective layer of **cartilage**. This is smooth and stops the bones rubbing on each other as they move. Around the joint is a lubricating liquid called **synovial fluid**. This 'oils' the joint. The knee joint is shown on the next page.

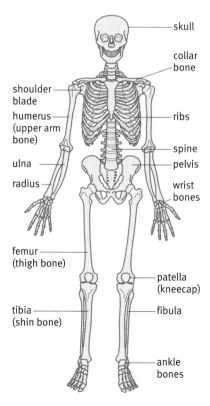

skull
collar bone
shoulder blade
humerus (upper arm bone)
ulna
radius
ribs
spine
pelvis
wrist bones
femur (thigh bone)
patella (kneecap)
tibia (shin bone)
fibula
ankle bones

Your bones contain calcium salts. Calcium minerals are important for good health.

Muscles move you

Muscles move the bones at a joint. A **tendon** connects a muscle to a bone. Muscles contract (get shorter) and pull on the bone. A muscle cannot push, it can only pull, so muscles work in pairs. One muscle pulls the bone one way, and the other pulls it back again. These pairs of muscles are called **antagonistic pairs**. On page 147 you can see that in your upper arm, your biceps and triceps muscles work as a pair.

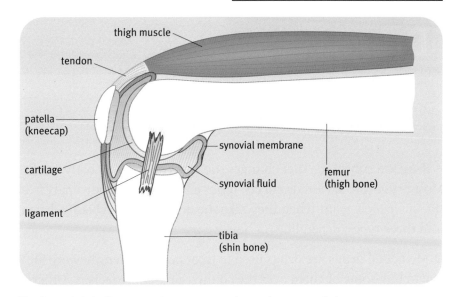

The knee joint allows you to move your lower leg up and down.

Exercising safely – taking care of joints

Exercise improves fitness, but you need to take care to exercise safely. It's important to warm up gradually or high-impact exercise can cause injury. Sometimes the stresses and strains of sport can damage a joint, such as:

o worn or torn cartilage
o stretched or torn ligament – a sprain
o pulled muscle – a strain
o a broken bone.

If a joint is badly damaged, doctors might decide to replace it. Replacement joints can give people a new lease of life after an accident, or to treat wear and tear on the joint.

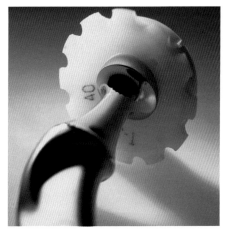

This replacement hip joint is hammered into the thigh bone, and the socket goes in the pelvis.

1 Copy and complete using words from the Language bank:

Your _____ supports your body and protects delicate organs. You can move your body at _____. Muscles work together in _____ to move the bones.

2 What problems can be caused by fatty deposits in the arteries? What can you do to try and avoid this happening?

3 Compare the movement at a ball-and-socket joint (hip or shoulder) with the movement at a hinge joint (knee or elbow). Which allows movement in all directions, and which just back and forward?

4 Draw a labelled diagram of a knee joint. Explain what each labelled structure does, and what problems the joint might suffer.

Language bank

angina
antagonistic pairs
bone
cartilage
high blood pressure
joints
ligament
muscle
skeleton
synovial fluid
tendon

O What effect do drugs have?
O Are we healthier than our great-grandparents were?

A **drug** is a substance that changes the way the mind or body works. Most drugs also have **side effects**, other effects on the body apart from the one people want when they take the drug. We can classify drugs into three types, depending on how we use them.

O **Medicinal drugs** are drugs used to treat people when they are ill. You can buy some of them over the counter at a pharmacist's. Others have more serious side effects and need to be prescribed by a doctor. Medicinal drugs are legal – it is not against the law to take them.

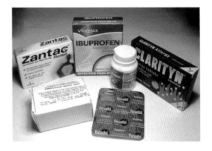

Medicinal drug	Effects	Side effects
paracetamol	painkiller	overdose can damage the liver
penicillin	antibiotic	some people are allergic to it

O **Recreational drugs** are substances that people take because they like the effect the drug has, not because they are ill. These drugs are sometimes called social drugs. Some of them are legal, such as caffeine in coffee and tea, nicotine in cigarettes, and alcohol.

Recreational drug	Effects	Side effects
caffeine	**stimulant:** speeds up the heart and makes you feel awake	can affect the heart and cause sleep problems
nicotine	stimulant	addictive, causes circulatory diseases

O **Illegal drugs** are drugs used for recreation, but it is against the law to take them. The reason why the drugs are illegal is because they are harmful. They have serious side effects, and many are highly addictive.

Illegal drug	Effects	Side effects
amphetamines (speed, whiz, uppers)	stimulant, makes you feel energetic and confident	addictive, body temperature rises, risk of dehydration
cannabis (dope, grass, marijuana)	depressant, can cause **hallucinations** (seeing things that are not really there)	feeling that everyone is against you, smoking-related illnesses and memory loss
ecstasy (E)	stimulant, can cause hallucinations	nausea, dehydration, overheating, possible brain and liver damage

Heroin comes from poppy seeds. It is illegal because it is addictive and harmful.

Classifying illegal drugs

It is extremely dangerous to take illegal drugs. The side effects are serious and the drugs are highly addictive, ruining people's lives. All an addict can focus on is how to get more drug, and this often means stealing and ending up homeless. The effects of the drug and an addict's lifestyle can lead to a very unpleasant death.

In the UK there are three classes of illegal drug. Fines and prison sentences for possessing, using or supplying these drugs depend on how dangerous the drug is.

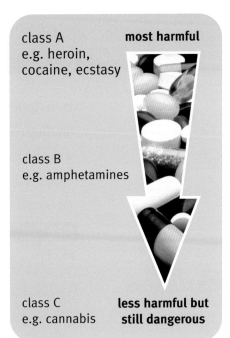

class A
e.g. heroin,
cocaine, ecstasy
most harmful

class B
e.g. amphetamines

class C
e.g. cannabis
**less harmful but
still dangerous**

Guess what?

The drug amphetamine sulphate (speed) is one of the dirtiest drugs around. A small amount of the pure drug is mixed (cut) with other materials including talcum powder, toilet cleaning powder and animal worming tablets.

A DIFFERENT WAY OF LIFE

YOUR GREAT-GRANDPARENTS probably grew up in the early twentieth century. At that time health care was not freely available as it is now, and conditions at work could be tough and dangerous. There were no laws about air pollution and many people suffered health problems. Smoking was popular – it was even seen as positively healthy. People did not have the great variety of food we eat today.

Today people on average live longer. Life is easier as most people use labour-saving devices such as cars and washing machines. We enjoy better food, free health care and modern medicine, and there are laws to protect our health and safety. However, teenagers are becoming less active and obesity is on the increase.

1 Copy and complete using words from the Language bank:

There are three types of drug: recreational drugs, _____ and _____. Some drugs have dangerous _____ and that is why they are illegal.

2 A new drug for indigestion is being developed, but researchers are worried about possible serious side effects. They could just give the drug to anyone with indigestion and see what happens. Explain whether you think this is a good idea.

3 Do you think teenagers today are healthier than their great-grandparents? Find out more about life 100 years ago. Make a list of lifestyle changes that would improve teenage health and make the most of our advantages.

Language bank

addictive
amphetamines
cannabis
ecstasy
hallucinations
heroin
illegal drugs
medicinal drugs
recreational drugs
side effects
stimulant

Checkpoint

1 Respiration

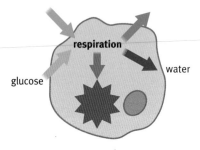

respiration

glucose

water

a Copy the diagram and write each label below in the correct place.

energy

carbon dioxide

oxygen

b Match each organ system with its correct role in helping cells carry out respiration.

Organ systems

circulatory system

respiratory system

digestive system

Roles

exchanging oxygen and carbon dioxide

providing glucose

transporting materials to and from cells

2 True or false?

Decide whether the following statements about a balanced diet are true. Write down the true ones. Correct the false ones before you write them down.

a A deficiency disease results if a nutrient is missing from the diet.

b Rickets is caused by a deficiency in vitamin C.

c Scurvy is caused by a deficiency in vitamin C.

d Obesity is a deficiency disease.

3 Dangerous drugs

Copy the following table. Choose the correct description to complete it for each drug.

Descriptions

an illegal drug which is a stimulant

a medicinal drug which is a painkiller

a legal addictive drug found in tobacco

an illegal drug which is a depressant

Drug	Description	Side effects
nicotine		diseases of the circulatory system
paracetamol		overdose can damage the liver
cannabis		smoking-related illnesses and memory loss
ecstasy		dehydration and damage to the brain and liver

4 Choose the answer

Copy and complete the following sentences, choosing the correct words.

During exercise, your heart beats **faster** / **slower** and your breathing rate is **faster** / **slower**.

If you exercise regularly, your heart gets **bigger** / **smaller** and your heart rate is **faster** / **slower**.

If you exercise regularly, your heart rate will recover more **quickly** / **slowly**.

5 Diseased parts

Match up the beginnings and endings to write four sentences.

Beginnings

Smoking

Fatty deposits in the arteries

Exercising without warming up

Too much salt

Endings

in the diet contributes to high blood pressure.

can cause bronchitis, emphysema, heart attacks, strokes and cancer.

could damage the joints or muscles.

may result from eating too much fatty food.

Plants and photosynthesis

Before starting this unit, you should already be familiar with these ideas from earlier work.

○ Like animals, green plants have organs such as leaves, roots and stems. Which part of a green plant carries out sexual reproduction?
○ Plants need light and water to grow well.
○ Plants make new materials in their leaves.
○ In the test for starch, iodine turns black showing starch is present. Can you think of some foods that have a positive starch test?
○ Respiration releases carbon dioxide. Which gas does it use?

You will meet these key ideas as you work through this unit. Have a quick look now, and at the end of the unit read them through slowly.

○ Green plants make their food by **photosynthesis**. This process does not just keep plants alive. It is also the basis of animal life, providing animals with their food and oxygen.
○ Photosynthesis takes carbon dioxide and water and converts them to glucose and oxygen. The process needs chlorophyll. The carbon dioxide for photosynthesis comes from the air, and the water comes from the soil.
○ Photosynthesis also needs energy from sunlight. Photosynthesis transforms the energy in sunlight into energy in glucose.
○ **Leaves** are adapted for photosynthesis. They are often broad and flat to catch lots of sunlight. Inside leaves are column-shaped cells with lots of chloroplasts, arranged near the top of the leaf to catch the light.
○ **Roots** are adapted for taking in water, as well as anchoring the plant in the soil. Roots are branched and spreading. They have tiny **root hairs** to help them absorb lots of water.
○ The glucose made by a plant is converted to lots of other materials in the plant, such as proteins and fats. All the materials in a plant make up its **biomass**. Plants need extra elements to convert glucose to all these other chemicals. They get these elements from **minerals** in the soil.
○ If minerals are in short supply in the soil, farmers use fertilisers.
○ Plants carry out respiration as well as photosynthesis. The word equations for these processes will show you how they are linked.

○ **How do plants grow?**

Look at the shoot growing from an acorn, pushing through the soil. Fifty years later it could be 30 metres tall.

The tree contains lots of material which was not in the seed. It has gained **biomass**. Biomass is the mass of all the material in a plant or animal, not including the water. Where do plants get the food and energy they need to grow and make all this biomass?

Food-making machines

Green plants do not absorb all their food from the soil, and they do not eat it like we do. Plants make or synthesise their food in a reaction called **photosynthesis**. This uses the raw materials carbon dioxide from the air and water from the soil to make **glucose**. Oxygen gas is given off as a by-product.

Photosynthesis needs energy, and this comes from sunlight. Plant cells have **chloroplasts** which contain a green pigment called **chlorophyll**. This absorbs sunlight energy and drives photosynthesis.

sunlight

carbon dioxide from the air

oxygen

photosynthesis

glucose

water from the soil

Don't plants get their food from the soil?

If the biomass of all the trees in a forest came from the soil, there would be no soil left! Plants make their own food.

Here is the word equation for photosynthesis:

sunlight
carbon dioxide + water ⟶ glucose + oxygen
chlorophyll

Investigating photosynthesis

Four things are needed for photosynthesis: carbon dioxide, water, sunlight and chlorophyll. On the next page are some experiments to show that photosynthesis needs these four things.

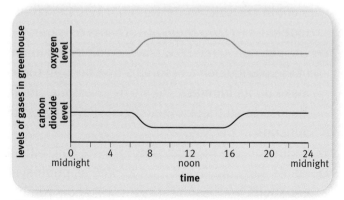

Photosynthesis uses carbon dioxide and produces oxygen during the day. What time do you think it went dark?

We can tell when a plant has been photosynthesising, because it will have starch in its leaves. Glucose is converted to starch and stored. We can test leaves for starch using iodine.

1 Plants need carbon dioxide for photosynthesis
Take two plants and set them up like this. After a few days carry out a starch test. Plant 1 has no starch in its leaves, showing it has not been photosynthesising. Plant 2 has starch in it leaves. Carbon dioxide is needed for photosynthesis.

plant 1
no carbon dioxide
clear polyethene bag
plant 2
solution of sodium hydrogen-carbonate – gives off carbon dioxide
soda lime – removes carbon dioxide
plenty of carbon dioxide

2 Plants need water for photosynthesis
If you don't water a plant, it wilts and dies. Photosynthesis and other biological processes need water.

3 Plants need sunlight for photosynthesis
If you leave a plant in a dark place for a few days, its leaves turn pale. If you test a leaf with iodine, you will find no starch. The plant cannot make its food without sunlight.

The bean plant on the left has been grown in the dark. Its leaves are pale and it has grown spindly trying to find light.

4 Plants need chlorophyll for photosynthesis
Some geranium leaves are **variegated** – they are green and white. The green parts contain chlorophyll, but the white parts do not. If you test a variegated leaf with iodine, you will find that the white parts do not contain starch but the green parts do. Chlorophyll is needed for photosynthesis.

The black areas show that starch was present only where the leaf was green.

1 Copy and complete using words from the Language bank:

Plants make their own _____ using a reaction called _____. The reactants are _____ from the air and _____ from the soil. _____ and _____ are also needed. The reaction makes _____ and oxygen.

2 Mrs Green put her yucca plant in the airing cupboard to keep it warm while she went on holiday. Explain what might happen to the plant.

3 Correct these statements:
   a Hydrogen is essential for photosynthesis.
   b Photosynthesis happens quickest at night.
   c Photosynthesis produces a sugar called sweetose.

4 When investigating photosynthesis, we change only one factor at a time. Explain why.

Language bank

biomass
carbon dioxide
chlorophyll
energy
food
glucose
oxygen
photosynthesis
starch
sunlight
water

- What is the role of the leaf in photosynthesis?
- What happens to the glucose produced in leaves?

The **leaf** is the plant organ where photosynthesis happens. Leaves are **adapted** for photosynthesis. They are broad and flat so they catch lots of sunlight. Cells in the leaf contain chloroplasts, which have chlorophyll.

The leaf – designed for photosynthesis

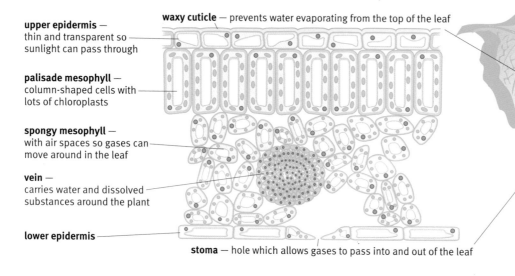

waxy cuticle — prevents water evaporating from the top of the leaf

upper epidermis — thin and transparent so sunlight can pass through

palisade mesophyll — column-shaped cells with lots of chloroplasts

spongy mesophyll — with air spaces so gases can move around in the leaf

vein — carries water and dissolved substances around the plant

lower epidermis

stoma — hole which allows gases to pass into and out of the leaf

The palisade mesophyll cells are where most photosynthesis happens. They are tightly packed together at the top of the leaf so they can trap lots of sunlight.

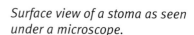

Surface view of a stoma as seen under a microscope.

On the underside of the leaf are holes called **stomata**. They allow carbon dioxide to enter the leaf for photosynthesis, and oxygen to leave.

What happens to the glucose?

Plants make glucose and oxygen by photosynthesis. But like all living things, plant cells also respire to release energy for the life processes. So plants use some of the glucose and oxygen they make for respiration.

Photosynthesis:

$$\text{carbon dioxide} + \text{water} \xrightarrow[\text{chlorophyll}]{\text{sunlight}} \text{glucose} + \text{oxygen}$$

Respiration:

glucose **+ oxygen** ⟶ **carbon dioxide +** water   **energy released**

When it's sunny and lots of photosynthesis is happening, the plant makes more glucose than it needs for respiration. So there is plenty of glucose left over. This is converted to other substances in the plant. Some glucose is changed to starch and stored for later. Some glucose is changed to other chemicals such as proteins, sugars and fats. All these chemicals from photosynthesis make up the biomass of the plant.

Do plants only respire at night?

No, plants respire all the time. During the day they use the oxygen they make from photosynthesis. At night they need to take in oxygen for respiration.

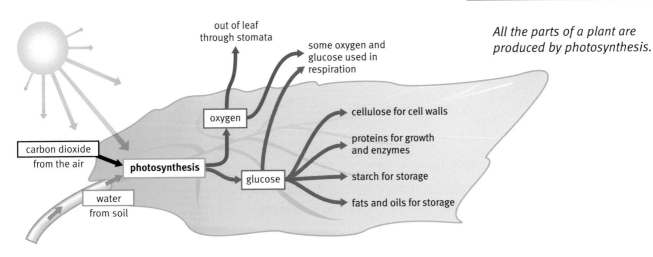

out of leaf
through stomata

some oxygen and
glucose used in
respiration

oxygen

cellulose for cell walls

proteins for growth
and enzymes

carbon dioxide
from the air

photosynthesis

glucose

starch for storage

fats and oils for storage

water
from soil

*All the parts of a plant are
produced by photosynthesis.*

Investigating the rate of photosynthesis

Pondweed is a plant that grows
in water. As it photosynthesises,
bubbles form in the water. You
can collect the gas and test it
with a glowing splint. It relights
the splint, showing that the plant
is producing oxygen.

If you measure the volume of
oxygen given off each minute,
this gives an idea of the rate of
photosynthesis – how fast it is
happening. You can change the
light intensity by moving the light further away from the test tube, or
change the temperature of the water. You can see how these changes
affect the rate of photosynthesis.

light source

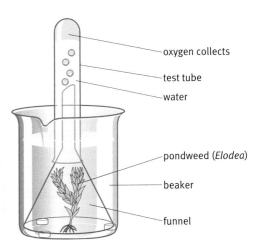

oxygen collects

test tube

water

pondweed (*Elodea*)

beaker

funnel

*Oxygen collects as the pondweed
photosynthesises. You can
measure the volume of gas
produced in a certain time.*

1 Copy and complete using words from the Language bank:

The leaf is the plant _____ where photosynthesis takes place. Extra
glucose from photosynthesis is stored as _____. Plant cells need
oxygen for _____, just like animal cells. During the day they use the
_____ from photosynthesis. At night they get it from the air.

2 Make a table of features of the leaf, showing how each one helps the
leaf to carry out photosynthesis.

3 a Mary collected 2 cm³ of oxygen from pondweed in 30 seconds.
She moved the light closer and collected 2.5 cm³ of oxygen in
30 seconds. Explain what this shows.

b Look at page 35. If Mary added sodium hydrogencarbonate to the
pondweed water, what might she expect to see?

4 Find out what people thought about plant growth before they knew
about carbon dioxide. Try searching for these names: van Helmont
and Priestley.

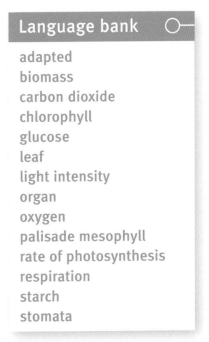

Language bank

adapted
biomass
carbon dioxide
chlorophyll
glucose
leaf
light intensity
organ
oxygen
palisade mesophyll
rate of photosynthesis
respiration
starch
stomata

37

The roots

○ **What is the role of the root in photosynthesis?**

The role of the root

Plants need a good supply of carbon dioxide and water for photosynthesis. You have seen how carbon dioxide from the air enters the leaf. What about the water? Water enters the plant from the soil, through the **roots**. The roots are plant organs that are adapted to:

○ absorb water and dissolved minerals
○ anchor the plant in the soil.

Roots spread through the soil, branching again and again. They are covered in cells called **root hair cells**. These have tiny threads called **root hairs** which give the roots a large surface area. The cells have very thin walls. All these adaptations help the roots to take in lots of water.

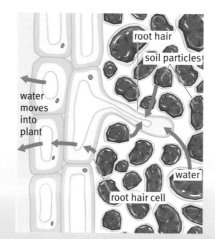

The root hair cells increase the surface area for taking in water.

A garlic clove suspended in water grows a lot of roots.

Why plants need water

A raw material for photosynthesis

$$\text{carbon dioxide} + \text{water} \xrightarrow[\text{chlorophyll}]{\text{sunlight}} \text{glucose} + \text{oxygen}$$

To provide essential elements

As well as carbon dioxide and water, plants also need particular elements so they can make biomass. They need nitrogen to make proteins for growth, phosphorus for growing good roots, and potassium for flowers and fruit. Dissolved in soil water are **mineral salts** which provide these elements for plants. Nitrates provide nitrogen, phosphates provide phosphorus, and potassium salts provide potassium. These mineral salts pass into the plant through the roots, dissolved in water.

If a plant does not get enough dissolved minerals, it will not grow well. It will show deficiency symptoms:

○ Without nitrates plants grow slowly and their leaves are pale.
○ Without phosphorus the roots do not develop and the plant may turn purple.
○ Without potassium there are no flowers or fruit and the leaves are yellow or brown.

Farmers add **fertiliser** to soil to provide enough minerals for plants.

Guess what?

Like all other cells, the cells of the root need oxygen for respiration. They get this oxygen from air in the soil. If the soil becomes waterlogged the root cells cannot get enough oxygen and the plant may die.

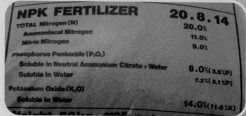

NPK FERTILIZER	20.8.14
TOTAL Nitrogen (N)	20.0%
Ammoniacal Nitrogen	11.0%
Nitric Nitrogen	9.0%
Phosphorus Pentoxide (P_2O_5)	
Soluble in Neutral Ammonium Citrate + Water	8.0% (3.5%P)
Soluble in Water	7.2% (3.1%P)
Potassium Oxide (K₂O)	
Soluble in Water	14.0% (11.6%K)

This fertiliser shows the relative amounts of N, P and K that it provides. What do you think these letters stand for?

To transport substances around the plant

Veins run all the way up the stem, connecting the roots to the leaves. Water and dissolved minerals are absorbed by the roots. They pass up the stem to the leaves in the veins. Water evaporates from the leaves, and more water is pulled up the veins to replace it (see opposite).

The veins contain xylem tubes and phloem tubes. The xylem tubes carry water and dissolved minerals up the plant. The phloem tubes carry substances made by the plant, such as sugars and plant hormones, around the plant to where they are needed.

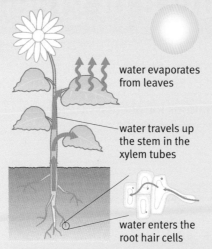

water evaporates from leaves

water travels up the stem in the xylem tubes

water enters the root hair cells

As water moves throughout the plant, it carries dissolved minerals with it.

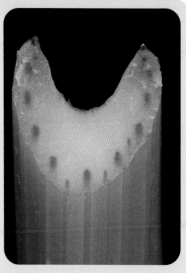

This celery has been standing in a dye solution, and you can see that the dye is only in the veins.

To keep the plant upright and cool

When plant cells have plenty of water, the cytoplasm presses against the cell wall. This pressure makes a plant rigid and keeps it standing up straight.

Water evaporates from the leaves, helping to keep the plant cool on hot days. This is like sweat evaporating from your skin. If a plant gets too hot, it cannot photosynthesise well.

For cell reactions and juicy fruits

There is lots of water in the cytoplasm of plant cells. The reactions of life happen dissolved in this water, so water is very important to a plant.

Water makes the plant's fruits nice and juicy.

Language bank

anchor
deficiency symptoms
evaporation
mineral salts
nitrogen, N
phloem
phosphorus, P
photosynthesis
potassium, K
respiration
root hair cells
roots
surface area
veins
water
xylem

1 Copy and complete using words from the Language bank:

   The roots are adapted to take in _____ which is used in _____. They also _____ the plant in the soil. Roots have a large _____ to absorb lots of water.

2 How do plants use water?

3 Why do farmers put fertiliser on their fields?

4 Andrew's house plants have purple younger leaves and their roots are underdeveloped. He thinks it is due to a lack of either nitrogen, phosphorus or potassium. How might he find out which?

○ **Why are green plants important in the environment?**

Look at the word and symbol equations for photosynthesis and respiration:

Photosynthesis (green plants):

$$carbon\ dioxide + water \xrightarrow[\text{chlorophyll}]{\text{sunlight}} glucose + oxygen$$

$$6CO_2 + 6H_2O \rightarrow C_6H_{12}O_6 + 6O_2$$

removes carbon dioxide from the air

Respiration (all living things):

$$glucose + oxygen \rightarrow carbon\ dioxide + water \qquad energy\ released$$

$$C_6H_{12}O_6 + 6O_2 \rightarrow 6CO_2 + 6H_2O$$

adds carbon dioxide to the air

Plants photosynthesise: they take carbon dioxide out of the air and produce glucose and oxygen. Plants and animals respire: they use glucose and oxygen and produce carbon dioxide. This makes plants vitally important to our existence. Without plants, animals would have no food and also no oxygen for respiration. We would all die.

Respiration depends on photosynthesis

Look at the illustrations. The first mouse has plenty of food and water, but once it has used up all the oxygen for respiration it will suffocate and die.

The second mouse has no oxygen worries. The plants remove the carbon dioxide it produces, and make oxygen for it to use. It needs lots of plants to do this. If we took some plants out, carbon dioxide gas would build up and oxygen levels would fall.

The Sun is the most important thing here as it provides energy for the whole system. Sunlight allows the plants to make food from carbon dioxide and water. Without the Sun we would all be in big trouble.

Upsetting the balance

If the amount of carbon dioxide used for photosynthesis in the world equals the amount of carbon dioxide produced by respiration, then things will remain in balance. But as you know, carbon dioxide is not made only by respiration. Combustion of fossil fuels also produces vast quantities of the gas. The oceans absorb carbon dioxide from the atmosphere and help keep things in balance. But if huge areas of the world's forests are removed every day, the amount of carbon dioxide in the air could rise and rise. You can read about global warming and the problems this causes on page 88–9.

Guess what?

Carnivores are meat eaters. But even carnivores depend on green plants, because the animals lower down the food chain are **herbivores** which eat plants.

Problems with deforestation

Removing trees on a large scale is known as **deforestation**. People living in the rainforests carry out deforestation for several reasons. They may want to use the wood for furniture or building materials. Or they may want to use the land to grow crops instead. In this case they may burn the trees, producing more carbon dioxide. Cutting down trees and burning them is known as slash and burn.

A deforested area of the Amazon rainforest. A third of all the world's species live and reproduce in the Amazon.

When forests are cut down, the mineral salts are easily washed out of the soil, making it harder for crops to grow there. Soil can be washed away, causing pollution and silting up rivers and lakes.

Forests provide habitats for many species of plants and animals. Some research suggests that thousands of species become extinct every year due to deforestation.

Conserving forests

For local farmers in the Amazon rainforests, clearing trees to farm the land is a way of life that has gone on for generations. Organisations are trying to educate these farmers to reduce the impact of deforestation.

○ They are encouraging farmers to plant crops which grow in the shade, like coffee and cocoa. The forest does not have to be completely cleared for this, so it can grow back quite quickly. They are encouraging the farmers not to burn the trees.

○ Large logging companies can be granted only limited contracts, and if they carry out selective logging many trees are left standing.

○ Some organisations have purchased large areas of rainforest. They plan to develop the forest in a sustainable way so the local people can live well without damaging nature.

○ Here we can help by using alternative materials to wood, recycling paper, buying products that use recycled materials, and supporting **aforestation**, which means planting new trees.

Guess what?

It is estimated that an area of 20 football pitches of forest is cut down every minute.

Growing seedlings for aforestation in Ecuador.

1 Copy and complete using words from the Language bank:

Photosynthesis removes _____ from the air and adds _____. Respiration removes _____ from the air and adds _____. If these processes are balanced, then the amount of carbon dioxide in the air will remain constant. Other reactions like _____ may upset the balance.

2 Draw a flow chart to show how carbon dioxide, water, light, the leaf and the roots are all needed for photosynthesis.

3 Use the information on this page and from other sources to discuss why the rainforests are important to the world.

Language bank

aforestation
carbon dioxide
combustion
deforestation
global warming
oxygen
photosynthesis
respiration
slash and burn
sustainable development

41

Checkpoint

1 The process of photosynthesis

Sketch this diagram. Choose the correct labels below for A to F.

glucose
sunlight
chlorophyll
water
oxygen
carbon dioxide

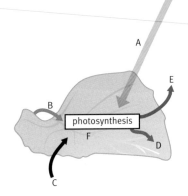

2 True or false?

Photosynthesis is vital to all life, both plant and animal. Choose the correct reasons for this below.

- It provides all animals with food.
- It uses water and stops it building up in the environment.
- It provides animals with somewhere to sleep.
- It transforms the energy in sunlight into energy in glucose.
- It provides animals and plants with carbon dioxide for respiration.
- It provides animals and plants with oxygen for respiration.
- It breaks nutrients up into smaller pieces that can be absorbed.

3 Adapted for photosynthesis

Copy the table below. Choose the correct function for each part of the leaf from the following list and write it in your table.

Part of leaf	Function
waxy cuticle	
palisade mesophyll	
spongy mesophyll	
stomata	
vein	

Functions

allow gases to pass in and out of the leaf
prevents water evaporating
has spaces for gases to move around the leaf
carries water and dissolved substances around the plant
photosynthesis

4 Getting roots right

Read the following sentences. Unscramble the bold words before you write them down.

Roots **hocran** the plant in the soil.
Roots have tiny **toro shari** to increase their surface area.
Roots are **darbench** and spreading to increase their surface area.
Roots take in **ratew** and mineral elements from the soil.
Root cells need oxygen so they can be damaged if the soil is **gatewordleg**.

5 Respiration and photosynthesis

a Copy and complete the following word equations. The same words are missing from each one. Label them to show which is respiration and which is photosynthesis.

Equation A

glucose + oxygen → _____ + water   energy released

Equation B

$$\overset{\text{sunlight}}{\underset{\text{chlorophyll}}{\text{_____ + water → glucose + oxygen}}}$$

b Which of the two equations is the source of biomass in a plant?

Plants for food

Before starting this unit, you should already be familiar with these ideas from earlier work.

- Plants and animals alike carry out the life processes, including movement, growth, reproduction and nutrition. How is nutrition different in animals and plants?
- Plants carry out photosynthesis to make glucose. They convert this glucose into many other materials, using elements such as nitrogen from the soil.
- All the **biomass** of a plant comes from photosynthesis.

You will meet these key ideas as you work through this unit. Have a quick look now, and at the end of the unit read them through slowly.

- Photosynthesis provides not only all our food, but also the oxygen we need for respiration.
- Food webs and pyramids of numbers show how all the organisms in a habitat depend on each other. The source of all their food is photosynthesis, powered by sunlight energy.
- Pyramids of numbers have a limited number of levels. This is because of energy losses from the food chain. Energy enters as sunlight energy, which is transferred to chemical energy in the biomass of plants.
- Each animal uses some energy for the life processes, and loses some heat energy to the surroundings. The more links there are in a food chain, the more energy has been lost.
- Farmers change the conditions in a habitat so they can produce the maximum amount of food from it. This changes the balance in the environment, affecting all the other animals and plants.
- Farmers put fertilisers on the soil so the plants have enough minerals. In a greenhouse, growers can control the amount of water, the temperature, carbon dioxide levels and light levels.
- Weeds and pests reduce the **crop yield**. They compete with humans for food. Farmers use many methods to reduce this competition, including weedkillers and pesticides.
- These chemicals can affect other organisms in the habitat. Poisonous **toxins** can build up in the food web.

The food we eat

○ Where does our food come from?

Remember **food chains**? They show which organisms eat which other organisms, and the **feeding relationships** between them. Food chains link up to make **food webs**, showing all the feeding relationships in a habitat.

Here are some food chains for this meal. The arrows show the direction of energy flow through each food chain:

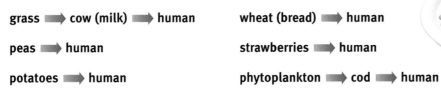

grass ⟹ cow (milk) ⟹ human   wheat (bread) ⟹ human

peas ⟹ human   strawberries ⟹ human

potatoes ⟹ human   phytoplankton ⟹ cod ⟹ human

The Sun is the energy source for all food chains, and they all start with green plants. These are **producers** – they can photosynthesise using light energy, producing glucose from carbon dioxide and water. The animals are **consumers**. The first consumers, that eat green plants, are called **primary consumers**, and the next ones are **secondary consumers**.

| phytoplankton | → | cod | → | human |

Producer – takes the raw materials carbon dioxide and water, and transfers the Sun's energy into energy in glucose. This might be stored as starch or converted to proteins, fats and other substances.

Primary consumer – transfers energy from glucose and starch in the plant into energy within the fish. It uses this energy to move, grow and live.

Secondary consumer – transfers energy from the fish's body into energy in the human body. Humans in turn use this energy to move, grow and live.

The diagram shows how energy from the Sun is transferred through a food chain. But each time energy is transferred from one link to the next, some energy leaves the food chain. The phytoplankton uses some energy for its life processes. Some parts of the phytoplankton will not be digested, so its body does not all end up as part of the cod's body. In the same way, the cod uses energy for the life processes, and some of its body such as the bones and scales are not good to eat. So not all the energy in the cod gets transferred to the human.

Remember **pyramids of biomass**? The bars get smaller towards the top because some energy leaves the food chain at each stage.

milk

strawberries

bread

cod

ice cream

peas

potatoes

Guess what?

Plankton is microscopic organisms which float in the sea or in fresh water. Phytoplankton is plant plankton and zooplankton is animal plankton.

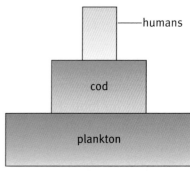

humans

cod

plankton

The width of each bar shows the biomass. The biomass of cod is smaller than the biomass of plankton.

The top bar in this pyramid represents those humans who eat cod, not the entire human race. Humans are **omnivores** – we eat lots of different foods so we feature in many food chains, at the top of a very complex food web.

Plants as food

You know that some plants are good to eat. Different parts of the plant provide us with different kinds of food. Plants carry out lots of photosynthesis in the summer when there is plenty of sunlight, and convert the glucose they make into other types of biomass to store.

Some plants store starch in their bulbs or roots, ready to use for growth in the spring. These bulbs and roots are a good source of carbohydrate for us. Some plants store starch and oils in their seeds, ready for when the seed starts to germinate and produce shoots and roots. These seeds are also good food sources for us.

Many plant leaves contain vitamins and minerals. Fruits are often rich in sugar and contain a lot of water, so taste sweet and refreshing. This helps the plants disperse their seeds, as animals eat the fruit and spread the seeds in their faeces. But some parts of plants are not good to eat, and some such as rhubarb leaves are poisonous.

It is energy from the Sun that allows you to eat, move and carry out all your life processes.

Plant	Part that we eat
onion	bulb
carrot	root
celery	stem
potato	root
apple	fruit
soya bean	seeds
lettuce	leaves

The table shows some plant parts that we use as food sources.

You can show that potatoes contain starch using iodine solution. The black colour shows starch is present.

Language bank

bulbs
consumers
energy
feeding relationships
food chains and webs
fruits
leaves
photosynthesis
producers
pyramids of biomass
roots
seeds
starch
stems
Sun

1 Copy and complete using words from the Language bank:

   We use _____ to show which organisms eat which others. They show the flow of _____ and the feeding relationships in a habitat.

2 Using the following terms, draw a concept map about our food: producer, consumer, food web, photosynthesis, energy, Sun, glucose, starch, root, leaf, stem.

3 Which part of a plant do we eat when we have:
   a a Brussels sprout   **b** a tomato   **c** a hazelnut?

4 If everyone hunted their own food, around 6 billion fewer people would be able to live on this planet. Find out how food production is organised in this country, and write some short notes.

45

Are fertilisers plant food?

No, plants make their own food. Fertilisers provide mineral salts.

○ How do fertilisers affect plant growth?
○ How does competition with other plants affect plant growth?

Gardeners and farmers know that plants need a small but regular supply of mineral salts for healthy growth. Plants take in mineral salts from the soil, so eventually the soil will run out of them. This is why fertilisers are added to the soil.

Minerals for healthy plant growth

The three most important elements needed for good plant growth are nitrogen, phosphorus and potassium (NPK). In the fertiliser on page 38 there is a high proportion of nitrogen. A farmer might use this if the crop especially needs nitrogen. Different plants need a different balance of minerals.

○ Nitrogen is needed for healthy leaves and by plants which have a lot of greenery. Lawns need plenty of nitrogen.
○ Phosphorus is needed for healthy roots and flowers. Bulbs or shrubs might be given a higher proportion of phosphorus.
○ Potassium is needed for good general health and for flowers and fruit. It helps plants withstand extremes of heat and cold and fight disease.

Without nitrogen the leaves turn pale.

Without phosphorous the leaves turn purple.

The table shows some other minerals that plants need, which fertilisers can supply.

Mineral containing	Needed for
calcium	growth of young roots and shoots
magnesium	making chlorophyll and healthy seed formation
sulphur	making proteins and for healthy colour
iron	making chlorophyll, for photosynthesis and for healthy colour

Fertilisers are very expensive, so why do farmers use them? They help the crop grow well, so farmers can get the maximum **yield** from a piece of land. This means they can produce the largest possible amount of crop for every square metre of land.

Competition between plants

Organisms living in the same habitat compete with each other for resources from the environment. A **weed** is any plant growing where

Guess what?

Manure is a good natural fertiliser, so many gardeners put horse muck on their roses.

Weedkillers can be sprayed over a large area.

The weed wiper tackles tall weeds which brush against the wick.

we don't want it. A poppy might be a prized specimen in a garden, but in a cornfield it's a weed. Weeds compete with the crop for sunlight, for minerals in the soil and for water. Weeds can reduce the crop yield, costing the farmer money. To control weeds, farmers use **weedkillers**.

A weedkiller is also called a **herbicide**. Some herbicides simply need to touch the weed to kill it, while others work after being absorbed by the weeds' roots.

o **Non-selective herbicides** kill a wide range of plants. A farmer might use a non-selective herbicide on the soil to clear any weeds before planting a new crop.
o **Selective herbicides** target particular weeds. For example, some herbicides kill broad-leaved plants but not cereals, so when sprayed on a cornfield they will not affect the corn.

Weeds are part of the food web

Insects and other herbivores feed on weeds, and second consumers feed on these herbivores. Using weedkillers may affect many of the organisms in a food web.

Organic farmers choose not to use harmful weedkillers. They want to farm without damaging the land or the food chain. People can eat organic crops without worrying about eating weedkillers as well. Organic farmers may pull out weeds by hand, or leave them and accept that the crop yield will be lower.

Gardeners might target individual weeds.

This weed is a food source for the butterfly.

1 Copy and complete using words from the Language bank:

Plants need a small supply of _____ for healthy growth. Farmers add _____ to the soil to make sure their plants are healthy and to increase the _____. Weeds _____ with crop plants. Using _____ reduces the competition for resources in the field.

2 What other factors might affect the crop yield, apart from mineral salts and competition from weeds?

3 What might be the effect of using selective weedkillers on a food web? Give an example in your answer.

Language bank 　　○—

compete
crop yield
fertilisers
food web
mineral salts
nitrogen
non-selective herbicides
organic farmers
phosphorus
potassium
selective herbicides
weedkillers

47

Pests

What a pest!

Pests are a problem for farmers. A **pest** is an animal that eats the farmer's crop, so pests compete with humans for food. Insects, birds, mice and slugs are all pests that can destroy crops and reduce the crop yield. Farmers try to remove pests from the food web.

Dealing with pests

Pesticides are chemicals which kill pests. Some pesticides work when the pest touches or eats them. Others work from within the crop, so if the pest eats the crop it dies.

Pesticides contain poisonous substances called **toxins**. They can be hazardous to other organisms in a food web, not just the pests, so they need to be used with care. For example, slug pellets are dangerous to dogs, cats and small children who might eat them.

Slug pellets contain chemicals that kill slugs and snails, removing them from the food web.

Pest		How they compete with humans	Control methods
field mouse		eats crops such as wheat and barley	traps, cats, poison
cabbage white caterpillar		larvae eat crops such as cabbages and cauliflowers	microbial insecticide (contains bacteria which kill the larvae)
aphids		feed on many crops including tomatoes and gardeners' roses	pesticides, natural predators such as ladybirds
slugs and snails		eat the leaves of many plants including lettuces, strawberries and tomatoes	slug pellets, removing by hand, beer traps; deterrents such as eggshells on the ground, copper foil barriers

Pesticides and feeding relationships

Look at this food chain:

rose → aphid → ladybird → robin

Aphids may destroy a market gardener's rose crop. The pyramids of numbers show that spraying the roses with pesticide will reduce the number of aphids and give a better rose crop. But if the ladybirds cannot find other sources of food, their numbers will also go down, and this in turn will affect the robin population. Pest control has a knock-on effect on other organisms in the food web.

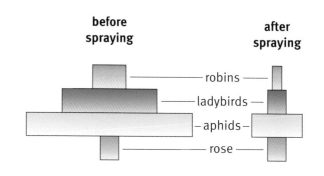

It is not just the pests that are affected by pesticides.

Pesticides and predators

Pesticides are expensive, especially when used on a large scale. But a good yield of undamaged crop will sell for a better price. Some pesticides contain toxins which are biodegradable – they are broken down naturally and do not get passed on in the food chain. However, other toxins do not break down.

There are alternatives to using pesticides. Organic farmers try to control pests without using harmful toxins. For example, snails and slugs have natural predators like birds and beetles. If these are encouraged, they will help control the pests and prevent them from destroying crops. Many gardeners use beer traps to trap slugs as these are more environmentally friendly than slug pellets.

Some varieties of plant have been developed which are resistant to aphids. This means we do not have to destroy the aphids. There are environmentally friendly insecticides such as spraying with soap solution, which is also cheaper than traditional insecticides.

Guess what?

A chemical with '...-icide' in its name gets rid of something. Herbicides get rid of herbs (plants), insecticides kill insects, rodenticides kill rodents and molluscicides kill molluscs (slugs and snails). What do you think a fungicide does?

Some people think using pesticides makes the problem worse. Slug pellets kill birds as well as slugs and snails. When the next snail invasion comes, there will be no birds to eat them, and the problem could be worse than before.

1 Copy and complete using words from the Language bank:

   _____ are animals that compete with humans. We use chemicals called _____ to control pests. They contain poisons, called _____. An alternative approach is to encourage natural _____ to eat the pests.

2 Pests are part of the natural food web. Why are they a problem?

3 How might using pesticides make the problem worse? Use the word 'predators' in your answer.

Language bank

aphids
crop yield
insecticides
pesticides
pests
predators
toxins

Toxins and food chains

○ How do pests affect plant growth?

In the 1950s, huge areas of land were sprayed with DDT.

Some pesticides are made from natural minerals or plant extracts. For example, sulphur and copper sulphate are used to treat fungal diseases on plants such as apple trees. Many modern pesticides are synthetic, made from chemicals from crude oil. Some of them contain chlorine and phosphorus. This makes them very effective at killing pests, but they can harm useful organisms as well. Pesticides containing toxins that do not break down naturally are particularly harmful.

The DDT solution

DDT is a very effective insecticide. After the Second World War, DDT was used to help control malaria and typhus. These are both serious diseases which are spread by insects. Malaria is carried by mosquitoes, and typhus by head lice. The DDT insecticide was sprayed over large areas to kill these insects, preventing the diseases from spreading. The table shows that it was successful.

	Number of cases of malaria	Number of deaths
before DDT	75 million	1 million
after DDT	5 million	5000

DDT had a massive effect on malaria cases and deaths in India. It was also used in many other countries.

DDT was also used as an insecticide in the UK and elsewhere to treat insect pests on crops. It was cheap and did not seem toxic to mammals. People thought it was an excellent insecticide.

The DDT problem

There were two problems with this widespread use of DDT. One was that some insects developed a resistance to DDT, so the spraying was no longer so effective. The second problem was that by the 1960s, scientists in the UK and elsewhere were finding high levels of DDT in top carnivores, animals at the top of the food web.

The reason for this is that the DDT toxin is persistent – it does not break down in the environment. All the DDT that was sprayed on the land stayed in the environment.

pondweed → tadpoles → minnows → perch → heron

Look at the food chain. The tadpoles ate the pondweed and plankton, eating a little DDT with each meal. Minnows ate the tadpoles, and all the DDT in the tadpoles passed to them. Remember that DDT does not

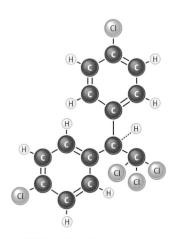

A pesky mosquito shouted out in pain ...

'A chemist has destroyed my brain!'

The cause of his sorrow was para-dichloro diphenyltrichloroethane

A DDT molecule.

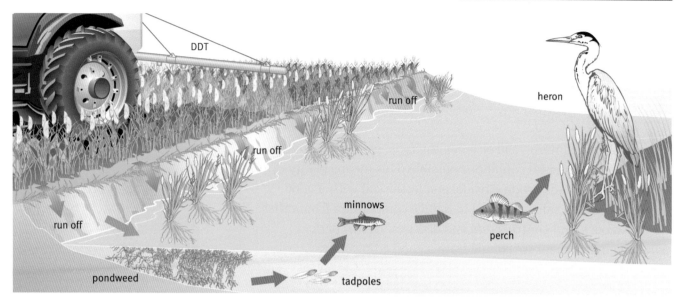

DDT was sprayed on crops to kill insect pests. Some DDT found its way into lakes, ponds and rivers. It was taken up by pondweed and entered the food chain.

break down – once eaten it stays in animals' bodies. Going along the food chain, the animals get bigger and they eat more and more of the smaller animals in their lifetime. Perch eat lots of minnows, and herons eat even more perch. The amount of DDT they take in gets more and more. The levels of DDT build up in the top carnivores. This is called **bioaccumulation**.

In herons, the amount of DDT per kilogram of body mass was larger than in the fish they ate. In some cases it was enough to kill the herons. At lower levels it made their eggshells very thin, so the eggs were crushed before they hatched.

Newer pesticides

DDT is banned in the UK and many other countries, to protect the environment. New chemicals have been produced which are similar to DDT, but which break down rapidly in the environment so do not pose a long-term danger to wildlife.

1 Copy and complete using words from the Language bank:

DDT is a _____ pesticide – it does not break down in the environment. It builds up in the food chain to high levels in the top _____. This building up is called _____.

2 a What does DDT stand for?
   b Explain how DDT helps in the fight against malaria.

3 DDT levels on crops were considered safe. Why did its use cause problems?

4 Some people argue that DDT has saved more human lives than it has killed wildlife. What do you think about this? Do you think DDT should be banned worldwide?

Guess what?

The name DDT comes from the old name DichloroDiphenyl-Trichloroethane. More correctly it should be called 1,1,1-trichloro-2,2-bis(4-chlorophenyl)ethane. It has been banned in many countries but is still used in some developing countries where malaria needs to be controlled.

Language bank

bioaccumulation
carnivores
DDT
insecticide
malaria
mosquitoes
persistent
resistance
toxin
typhus

A perfect growing environment

What is the perfect environment for growing plants?

To grow the maximum yield of a healthy crop, we need to give it the best environmental conditions possible. This means plenty of sunlight, water and carbon dioxide, and fertile soil with a good supply of mineral salts. We need to reduce competition from other plants (weeds), and control pests and diseases. In this perfect growing environment we should have a good crop yield.

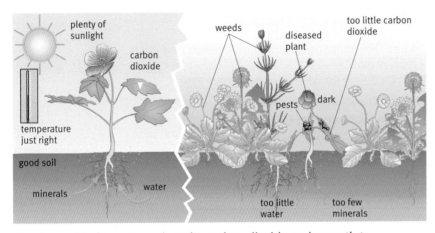

Plants need light, water, minerals, carbon dioxide and warmth to photosynthesise and grow well.

Growing grass in very dry conditions does not give the best results.

The natural environment

It is difficult to achieve perfect growing conditions for all plants. The conditions may be good in the middle of a field of wheat, for example, but around the edges some plants may be shaded from sunlight, or the drainage may be poor. People try and adapt the natural environmental conditions to allow crops to grow well.

In very dry areas, for example in Africa, natural water supplies are not enough. Irrigation ditches are dug to provide enough water for crops. Without irrigation these banana plants would not grow here.

In very wet areas, water-loving crops such as rice are grown. The flat terraces stop the water running off.

Greenhouses – being in control of the environment

Farmers have been improving the natural environment for centuries. Today's farmers have taken this to new levels. Growing crops in modern greenhouses allows them to control the temperature and carbon dioxide levels, and water and fertiliser are added automatically when needed.

Greenhouses extend the growing season, so crops can be grown and sold outside their traditional periods. It is easier to control pests and weeds, so crops develop in a regular and predictable way. Selective breeding provides crop plants with the best size, shape and taste to maximise sales.

This painting shows people controlling conditions to grow crops in the fifteenth century BC.

What cost?

There are a few disadvantages to this hi-tech farming. If pests get established, they spread very quickly in the warmth of a greenhouse. The greenhouses and automated systems are expensive, so the crop has to sell for a high price to justify this expense. Strawberries, flowers and tomatoes may be grown this way, but onions or potatoes sell for a lower price so it is not worth using hi-tech methods to grow them.

Artificial lights extend the daylight hours so lettuces grow faster.

Today controlling the growing environment is big business. Carbon dioxide is pumped in from cylinders to keep photosynthesis happening as fast as possible, and vents and heaters are automatically controlled to maintain a good temperature.

1 Copy and complete using words from the Language bank:

   To get the maximum crop yield, farmers try to provide ideal _____. They provide the raw materials for photosynthesis – light, water and _____. They also need to make sure there are enough _____ in the soil, and that the temperature is suitable for a high rate of _____.

2 What are the benefits of growing tomatoes in a greenhouse on a large scale?

3 Growing fruit developed by selective breeding in controlled conditions gives a consistent shape and taste. List some advantages and disadvantages of this.

4 Using the internet or other sources, find out how using greenhouses has extended the months of the year when we can buy fruits such as strawberries.

Language bank

carbon dioxide
crop yield
environmental conditions
greenhouse
irrigation
mineral salts
photosynthesis
sunlight
water

Checkpoint

1 Changing habitats

The following factors may all change a habitat. List them and classify each one as physical, chemical or biological.

- a swarm of bees
- spraying a field with pesticide
- lots of rain
- using fertiliser
- a cold spell
- a weed invasion

2 Limited by energy

Match up the beginnings and endings to make complete sentences.

Beginnings

A pyramid of numbers shows how many

The pyramid is never more than

This is because

Each animal takes in

It uses energy for the life processes,

There is much less energy

Endings

so not all the energy it took in is stored in its body.

organisms there are at each stage of a food chain.

four or five layers tall.

energy in its food.

at the top of the pyramid than at the bottom.

energy leaves the food chain at each stage.

3 True or false?

Decide whether the following statements are true. Write down the true ones. Correct the false ones before you write them down.

a Weeds are plants growing in the wrong place.

b Pests are animals that eat off people's plates.

c Weeds and pests both increase the crop yield.

d Slug pellets are an example of a pesticide.

e Pesticides and weedkillers may contain poisonous chemicals called toffees.

f If these toffees do not break down, they can build up in the food web.

4 Controlling conditions

Growers try to provide ideal conditions for photosynthesis in a greenhouse. List which conditions below they might control to maximise photosynthesis.

- hardwearing aluminium greenhouse frame
- water and mineral elements in the soil
- temperature and light
- average age of people that visit the greenhouse
- predators to eat insect pests
- carbon dioxide levels
- plenty of parking

5 A whole world view

Look at the following statements.

- All life depends on green plants.
- All the materials in the environment, such as carbon dioxide, oxygen and water, are recycled from one organism to another.
- Energy from sunlight flows through the environment.
- If we keep taking one resource out of the environment and not letting it be replaced, the environment will suffer and we will not be able to continue using it.

Design a big concept map, or a series of diagrams, or write a story to illustrate one or more of these points.

Reactions of metals and metal compounds

Before starting this unit, you should already be familiar with these ideas from earlier work.

- Metals are elements that have particular properties. One is that they conduct heat well. What other properties do metals share?
- In a chemical reaction, atoms join up in different ways.
- We use symbols with one or two letters to represent elements. What are the symbols for copper and carbon?
- When atoms are joined, we combine symbols into formulae. Can you remember what $CuSO_4$ stands for? (Hint: It's blue!)
- We describe chemical reactions in several ways – word equations, particle equations and symbol equations.
- If a gas is given off in a reaction, we can test it to see what it is. Can you remember the tests for hydrogen and carbon dioxide?
- In a neutralisation reaction, an acid is cancelled out. What sort of chemical neutralises an acid?

You will meet these key ideas as you work through this unit. Have a quick look now, and at the end of the unit read them through slowly.

- In a chemical reaction reactants are used up, and new products form. There might be a colour change, or bubbles. Another sign of a chemical reaction is energy transfer. Flashes, bangs and breaking glassware suggest that energy is given out!
- In chemical reactions, the atoms become rearranged. This can help you work out what the products might be. For example, if you heat copper in a gas jar of oxygen, and a reaction happens, what could be formed? There are not too many possibilities . . .
- You can use this idea of rearranged atoms to predict what might be formed in a reaction. Writing symbols and formulae helps.
- Because a chemical reaction is just a rearrangement of the atoms, the total mass of product is the same as the total mass of reactant. Mass is conserved in chemical reactions.
- Acids react with some metals. A salt is formed in these reactions, along with hydrogen gas.
- Acids also react with metal carbonates and with metal oxides to form a salt.

Metals

○ **Why are metals useful?**

Metals are elements

Do you remember how elements are organised in the periodic table? **Metals** are elements on the left-hand side of the table. Metals are usually shiny, hard and feel cold to the touch. **Non-metals** are elements on the right-hand side of the periodic table. Non-metals are usually dull, softer and do not feel cold to the touch.

H																	He
Li	Be											B	C	N	O	F	Ne
Na	Mg											Al	Si	P	S	Cl	Ar
K	Ca	Sc	Ti	V	Cr	Mn	Fe	Co	Ni	Cu	Zn	Ga	Ge	As	Se	Br	Kr
Rb	Sr	Y	Zr	Nb	Mo	Tc	Ru	Rh	Pd	Ag	Cd	In	Sn	Sb	Te	I	Xe
Cs	Ba	La	Hf	Ta	W	Re	Os	Ir	Pt	Au	Hg	Tl	Pb	Bi	Po	At	Rn
Fr	Ra																

Remember

Sulphur is a non-metal – it is an element on the right-hand side of the periodic table. Plastic, wood and concrete are not metals, but they are not non-metal elements either. They contain compounds, so you won't find them in the periodic table.

key
☐ metals   ☐ non-metals

Properties of metals and non-metals

Metals	Non-metals
conduct heat well	most conduct heat poorly
conduct electricity well	most conduct electricity poorly
strong	not as strong as metals
hard	softer than metals
very flexible – can be shaped and drawn into wires (ductile)	brittle – may shatter if you hit them
sonorous – make a noise like a bell when you hit them	not sonorous
shiny (though some may tarnish)	usually dull in appearance
high melting and boiling point	low melting and boiling point
dense	not very dense
iron, nickel and cobalt are magnetic	not magnetic

Some exceptions to the rule

Some metals and non-metals are a bit different. They do not follow the general properties shown in the table.

Graphite is a form of the non-metal element carbon. Graphite is shiny and it can conduct electricity, unusual for a non-metal. The layers of carbon atoms allow charge to pass through its structure.

 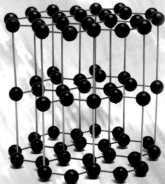

In graphite the carbon atoms are arranged in layers, as the model shows.

Diamond is also a form of carbon. In spite of being a non-metal diamond is the hardest natural material known. Its structure makes it very strong so it is used to tip drills and cutting tools.

Mercury is a metal element. It has a low melting point which makes it the only metal that is a liquid at room temperature.

Mercury is liquid metal.

How do we use metals?

The properties of metals make them useful in all sorts of ways.

Iron is strong and also quite cheap so we use it for bridges and buildings.

Where do metals come from?

Most metals are found in the ground not as the pure metal but combined in a compound – usually the oxide or sulphide. A mineral that contains a metal is called an **ore**. The table shows some ores.

Titanium is light but strong. It is also very expensive.

We dig out the ore and then extract the metal from it. A chemical reaction splits the metal from the ore. (You will find out more about this on page 77.) Metals are extracted from their ores in different ways.

o **Smelting**: roast (melt) the ore with carbon. Iron and zinc are extracted like this.

o **Electrolysis**: pass electricity through the melted ore or a solution of the ore. Aluminium is extracted by electrolysis.

o **Displacement**: react the metal ore with a more reactive metal. Titanium is extracted this way.

Metal ore	Main compound in ore
haematite	iron oxide
zinc blende	zinc sulphide
malachite	copper carbonate
galena	lead sulphide

Language bank

compounds
conductors
diamond
displacement
electrolysis
elements
graphite
mercury
metal
non-metal
ore
periodic table
properties
smelting

1 Copy and complete using words from the Language bank:

Metals are good _____ of heat and electricity while most non-metals are poor conductors. Metals are found on the left-hand side of the _____ while non-metals are found on the right.

2 Why is rubber neither a metal nor a non-metal?

3 Working in a group, each choose a different metal to research. Find out as much as you can about the metal:
  o What are its properties?
  o How is it used?

  o How do the properties make it useful in this way?
  o Where do we find the ore?

  Then build up a database of all the information the class found.

Metals and acids

O **What happens when metals react with acids?**

Bubbling away

Zinc reacts with both hydrochloric acid (left) and sulphuric acid (right).

Zinc, like many other metals, reacts with acids. Do you remember the signs that tell you a chemical reaction is happening? You know the zinc is reacting with the acid because:

o you can see bubbles of gas, a new material being formed
o the pieces of zinc become smaller
o the tube gets warm as heat is given out – energy is transferred
o if you evaporate the liquid left at the end of the reaction, it leaves a white solid called a salt, another new material.

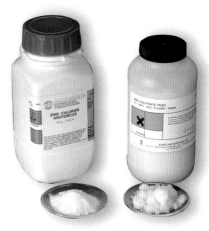

These salts are produced in the reactions.

What are the products?

Zinc reacts with acid to give off a gas and produce a salt. We can test the gas to see what it is. This reaction produces hydrogen gas.

Remember
The test for hydrogen is the squeaky pop test. The test for carbon dioxide is that it turns limewater cloudy.

Looking at the particles

When zinc reacts with acids the particles are rearranged:

Word equation: zinc + hydrochloric acid → zinc chloride + hydrogen
Symbol equation: Zn + $2HCl$ → $ZnCl_2$ + H_2

The zinc takes the place of the hydrogen in the acid to make the salt and hydrogen.

Word equation: zinc + sulphuric acid → zinc sulphate + hydrogen
Symbol equation: Zn + H_2SO_4 → $ZnSO_4$ + H_2

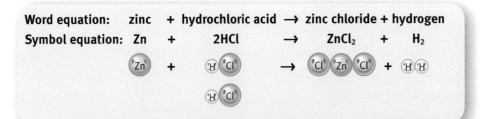

The zinc takes the place of the hydrogen in the acid to make the salt and hydrogen.

You can see how new materials are formed when the particles are rearranged in these chemical reactions. No new particles are added, so the total mass stays the same.

Patterns in the reactions

Many other metals react with acids like zinc does. But some metals do not react with acids.

calcium   magnesium   zinc   iron   lead   copper

The metals all react slightly differently. **Reactive** metals like calcium and magnesium bubble and fizz quite violently. Other metals like zinc and iron bubble away steadily. Metals that are less reactive than copper do not react with dilute acids.

We can write a general equation for the reaction of metals with acids:

metal + acid → salt + hydrogen

If a metal reacts with acid, this equation shows how it reacts.

A reactivity pattern

Some metals react more violently with acids than others. If we look at how metals react, we can list them in order of reactivity. We call this list the **reactivity series** or **activity series**. It is shown opposite.

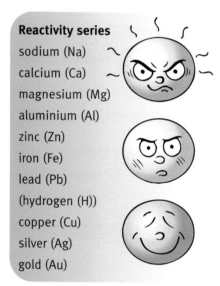

Reactivity series

sodium (Na)
calcium (Ca)
magnesium (Mg)
aluminium (Al)
zinc (Zn)
iron (Fe)
lead (Pb)
(hydrogen (H))
copper (Cu)
silver (Ag)
gold (Au)

If a metal is below hydrogen it will not react with dilute acids.

1 Copy and complete using words from the Language bank:

Some metals react with dilute acids like _____ and _____ . A _____ and hydrogen gas are formed. The most _____ metals react the most violently.

2 What is the test for hydrogen gas?

3 Write the general equation for a metal reacting with an acid.

4 Write a word equation for:
   a magnesium reacting with sulphuric acid
   b iron reacting with hydrochloric acid.

5 Predict whether silver will react with dilute hydrochloric acid. Explain your answer.

6 Write symbol equations for the reactions in question 4.

Language bank

activity series
general equation
hydrochloric acid
hydrogen
reactive
reactivity series
salt
sulphuric acid

Metal carbonates and acids

○ How do acids react with metal carbonates?

More bubbling reactions

You know that metals react with acids to make hydrogen gas and a salt. Metal carbonates also react with acids. This reaction produces carbon dioxide gas, a salt and water.

You can tell a reaction is happening because:

○ a gas is formed – a new material
○ the tube gets warm as heat is given out – energy is transferred
○ if you evaporate the liquid left at the end of the reaction, it leaves a salt – another new material.

Here is a word equation for the reaction:

magnesium carbonate + hydrochloric acid → magnesium chloride + carbon dioxide + water

We can write this general equation:

metal carbonate + acid → salt + carbon dioxide + water

Looking at the particles

Looking at the particles can help us work out what rearrangements are going on in the reaction. The table shows some particle pictures.

Substance	Particle picture	Formula
magnesium carbonate		$MgCO_3$
calcium carbonate		$CaCO_3$
sodium carbonate		Na_2CO_3
hydrochloric acid		HCl
sulphuric acid		H_2SO_4
nitric acid		HNO_3
water		H_2O
carbon dioxide		CO_2
hydrogen		H_2

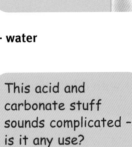

When magnesium carbonate reacts with sulphuric acid, the temperature rises.

This acid and carbonate stuff sounds complicated – is it any use?

Think about acid rain and limestone rock.

Acid rain is gradually destroying many ancient buildings made of limestone.

Let's take a closer look at this reaction:

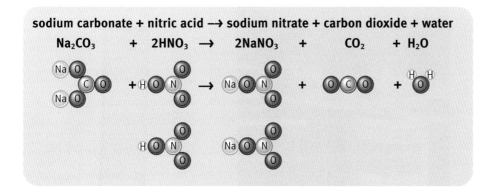

magnesium carbonate + sulphuric acid → magnesium sulphate + carbon dioxide + water

$MgCO_3$ + H_2SO_4 → $MgSO_4$ + CO_2 + H_2O

Naming salts

The first part of the name comes from the metal.

The acid you use gives the second part of the salt's name:
- sulphuric acid makes **sulphates**
- hydrochloric acid makes **chlorides**
- nitric acid makes **nitrates**.

This reaction forms the salt sodium nitrate:

sodium carbonate + nitric acid → sodium nitrate + carbon dioxide + water

Na_2CO_3 + $2HNO_3$ → $2NaNO_3$ + CO_2 + H_2O

Magnesium takes the place of hydrogen in the acid to form the salt magnesium sulphate. The carbonate part splits. It forms carbon dioxide, and loses an oxygen. Hydrogen joins with this oxygen to make water.

If we used calcium carbonate instead, the salt would be calcium nitrate. Carbon dioxide and water are also produced as before.

calcium carbonate + nitric acid → calcium nitrate + carbon dioxide + water

$CaCO_3$ + $2HNO_3$ → $Ca(NO_3)_2$ + CO_2 + H_2O

1 Copy and complete using words from the Language bank:

Acids react with _____ to produce a salt, _____ and water. When this happens the _____ rises, showing that energy is transferred.

2 Write a general equation showing how an acid reacts with a metal carbonate.

3 Using the pictures above, draw the particles in the reaction between zinc carbonate ($ZnCO_3$) and sulphuric acid.

Guess what?

Symbols and formulae are used all over the world, so scientists who speak different languages can understand each other's work.

Language bank

carbon dioxide
chloride
metal carbonates
nitrate
sulphate
temperature
water

Metal oxides and acids

○ What evidence is there of a chemical reaction between acids and metal oxides?

Another acid reaction

Copper oxide is a black powder. It reacts with sulphuric acid to make copper sulphate (a salt) and water.

sulphuric acid + copper oxide → copper sulphate + water

You can tell a reaction is happening because:
- the colour changes
- if you evaporate the blue solution you get crystals of copper sulphate, a new material
- the tube gets warm as heat is given out – energy is transferred.

There are no bubbles this time because no gas is produced.

Acids react with many metal oxides to make a salt and water. Here is the general equation for the reaction:

metal oxide + acid → salt + water

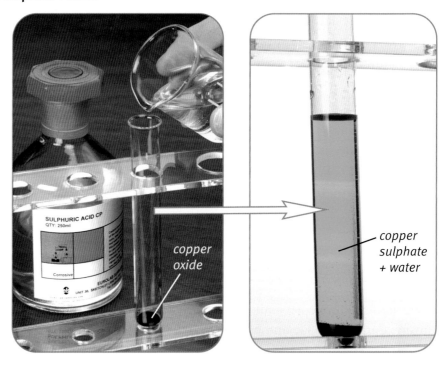

copper oxide

copper sulphate + water

Back to bases

A metal oxide is a **base**. Bases are substances that make a salt and water when they react with an acid. A soluble base is called an **alkali**. You have studied neutralisation reactions between alkalis and acids.

Looking at the particles

You can see how the particles are rearranged in this reaction:

copper oxide + sulphuric acid → copper sulphate + water

$$CuO \quad + \quad H_2SO_4 \quad \rightarrow \quad CuSO_4 \quad + \quad H_2O$$

The particles in copper oxide and sulphuric acid rearrange themselves to make copper sulphate and water in this neutralisation reaction.

Which salt?

o The metal in the oxide gives the first part of the salt's name.
o The acid gives the ending of the name.

Metal oxide	Acid		Salt	Other product
copper oxide	sulphuric acid	→	copper sulphate	water
copper oxide	hydrochloric acid	→	copper chloride	water
copper oxide	nitric acid	→	copper nitrate	water

Remembering neutralisation and pH

When an acid and an alkali react, they neutralise each other. The pH of the acid is 1 at the start, and the pH of the alkali is 14. At the end, the mixture is neutral, pH 7.

We can monitor the pH of a neutralisation reaction using a pH probe connected to a data logger. In this reaction acid is being added to alkali, so the pH is falling.

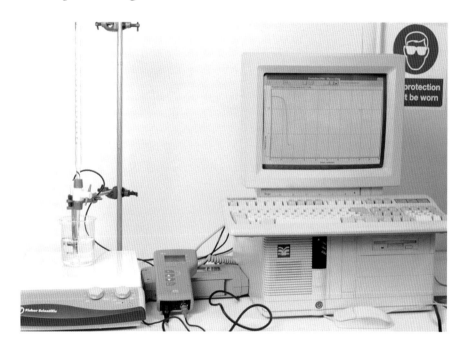

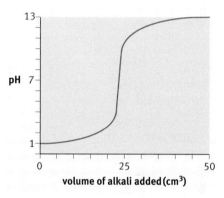

If alkali is added to acid, the pH rises instead.

1 Copy and complete using words from the Language bank:

An acid reacts with a _____ producing a _____ and water. The _____ is a new material, evidence that this is a _____ .

2 Write the general equation for a reaction between an acid and a metal oxide.

3 Write a word or symbol equation for the reaction of zinc oxide with:
   a hydrochloric acid   **b** sulphuric acid   **c** nitric acid.

   The formula for zinc oxide is ZnO.

Language bank

chemical reaction
data logger
metal oxide
neutralisation
pH
salt

What exactly is a salt?

What is a salt?

When you say 'salt' you probably think of the white stuff you put on chips. But that is just one salt, called common salt or sodium chloride. There are many others.

Salts are made in all these reactions:

metal + acid → salt + hydrogen
base + acid → salt + water
metal carbonate + acid → salt + carbon dioxide + water

Think of a salt as an acid with its hydrogen replaced by a metal:

potassium chloride

came from the metal, base or carbonate   *came from hydrochloric acid*

> **Remember**
> Bases include some metal oxides and metal hydroxides.
> An alkali is a soluble base.

These piles of sodium chloride have been extracted from the sea.

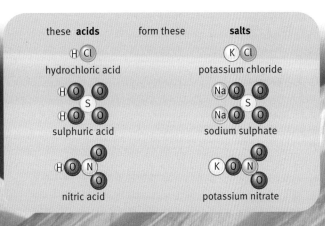

these **acids**   form these   **salts**

hydrochloric acid — potassium chloride
sulphuric acid — sodium sulphate
nitric acid — potassium nitrate

What would we do without salts?

Common salt (table salt) is sodium chloride. As well as making food tasty, sodium chloride is a major part of the world's oceans. In your body sodium chloride and other salts are kept in balance so your cells work properly.

Eating too much salt is bad for your health. Low sodium salt contains potassium chloride instead of sodium chloride for people who

want to reduce their sodium intake. Sea salt contains sodium iodide among other salts.

Plaster of Paris is used to set a broken arm. It is made from magnesium sulphate.

If you have greenfly on your prize plants you might spray them with greenfly killer containing copper sulphate.

pH tells us how acidic

You know that an acid has a pH lower than 7. An alkali has a pH higher than 7. A neutral solution has pH 7. Universal indicator tells us the pH of a solution.

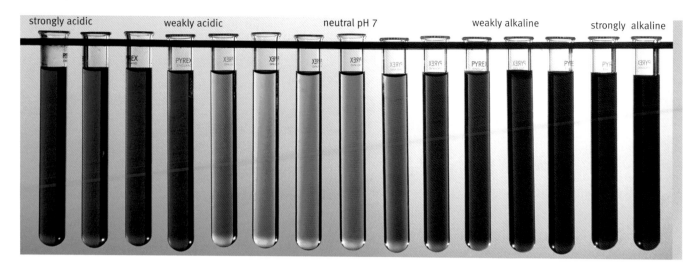

strongly acidic   weakly acidic   neutral pH 7   weakly alkaline   strongly alkaline

Litmus is another indicator. It tells whether a solution is acidic or alkaline, but not *how* acidic or alkaline.

Using universal indicator to check neutralisation

On page 63 you saw how we can use a data logger to follow a neutralisation reaction. Another method is to use universal indicator paper instead.

If you add alkali to acid bit by bit, you can test the pH after each addition to check whether it is neutral yet.

> **Remember**
> Acids and alkalis can be corrosive, harmful or irritant. Be careful with them and always wear goggles.

alkali —   glass rod —   glass rod —

acid —   UI paper —

add alkali to acid a little at a time

test pH of solution by dropping some onto universal indicator paper

When the solution is neutralised you can evaporate off the water to leave the salt behind.

1 Copy and complete using words from the Language bank:

When you add an alkali to an acid a _____ reaction happens. When it is complete the pH will be pH 7. You can check this with _____ .

2 **a** What is a salt?
  b Give one way of making a salt.

3 With an adult's permission, look through the labels on food and medicines at home. List all the salts that you find.

4 Find out what these salts are used for.
  a iron sulphate   **b** silver nitrate   **c** calcium phosphate

Language bank

acid
alkali
base
neutralisation
pH
salt
universal indicator

Checkpoint

1 Evidence of a chemical reaction

This word wall shows some signs that a chemical reaction is happening. Which of them are evidence of energy being transferred in the reaction?

2 Working it out

Look at these word equations. Copy and complete them. For each one, you can rearrange the reactants to work out what the products are.

a magnesium + oxygen $\longrightarrow$

b sodium carbonate + iron chloride $\longrightarrow$

c iron + sulphur $\longrightarrow$

d zinc + copper sulphate $\longrightarrow$

3 Word equations

Here is the general equation for an acid reacting with a metal:

metal + acid $\longrightarrow$ salt + hydrogen

Write word equations for the reactions at the top of the next column.

a magnesium with hydrochloric acid

b zinc with sulphuric acid

c calcium with nitric acid

d iron with sulphuric acid

4 Symbol equations

Here is the general equation for an acid reacting with a metal carbonate:

acid + metal carbonate $\longrightarrow$ salt + carbon dioxide + water

Copy and complete these symbol equations.

a $CuCO_3 + 2HCl \longrightarrow CuCl_2 + \underline{\quad\quad} + H_2O$

b $MgCO_3 + H_2SO_4 \longrightarrow \underline{\quad\quad} + CO_2 + H_2O$

c $FeCO_3 + \underline{\quad\quad} \longrightarrow FeSO_4 + CO_2 + \underline{\quad\quad}$

d $Na_2CO_3 + 2\underline{\quad\quad} \longrightarrow 2NaNO_3 + \underline{\quad\quad} + H_2O$

e $\underline{\quad\quad} + 2HNO_3 \longrightarrow Ca(NO_3)_2 + CO_2 + H_2O$

5 It's your choice

Copy and complete the following sentences, choosing the correct words.

A metal oxide reacts to neutralise **an acid / a base / a salt**.

The metal oxide is acting as **an acid / a base / a salt**.

The products of the reaction are a salt and **carbon dioxide / hydrogen / water**.

6 Missing materials

Copy and complete the table below.

Reactants		Salt formed	Other product/s formed
magnesium	nitric acid		hydrogen
	hydrochloric acid	copper chloride	carbon dioxide and water
zinc oxide	sulphuric acid	zinc sulphate	
iron		iron chloride	hydrogen

Patterns of reactivity

Before starting this unit, you should already be familiar with these ideas from earlier work.

○ Acids react with some metals, metal carbonates and metal oxides. A salt is formed in these reactions, along with other products. Can you write a word equation for the reaction between magnesium carbonate and hydrochloric acid?
○ Many metals react with oxygen in the air to form the metal oxide.

You will meet these key ideas as you work through this unit. Have a quick look now, and at the end of the unit read them through slowly.

○ Some metals react with oxygen very readily. The reaction starts as soon as you put the metal out in the air. You have to heat other metals to make them react, and there are some metals that will not react with oxygen at all. They stay shiny forever.
○ Metals can also react with water – some in spectacular fashion, others much more slowly.
○ This difference in speed of reaction is called **reactivity**. A metal that reacts readily, such as potassium, is a **reactive** metal. One that doesn't react at all, such as gold, is **unreactive**.
○ We can compare how readily different metals react with oxygen. Then we can write a list of the metals in order of reactivity. A list like this is called a **reactivity series**.
○ The reactivity series of metals reacting with water and with oxygen is the same as the series of metals reacting with acid.
○ We use metals in different ways according to how reactive they are. For example, using caesium to make knives and forks would make an explosive dinner party! We need a metal that won't react with air, water or dilute acids for long-lasting cutlery.
○ We can also predict how a particular metal might react by looking at its place in the reactivity series. If magnesium bubbles fast in hydrochloric acid, and we know that zinc is less reactive, we might expect zinc to react but not to bubble quite so fast.
○ A metal may react with the solution of a salt of another, less reactive metal. It takes the place of the less reactive metal in the compound. This is called **displacement**.

Dull or shiny?

○ Why do metals tarnish?

This gold necklace is about 2000 years old, but it looks almost new. The copper coin is the same age but it is no longer shiny like the new coins. It has become dull and **tarnished**. The metal has reacted with moisture and oxygen in the air over the years.

some react with acids **HARD** *conduct heat* **SHINY** *conduct electricity*

Do you remember the properties of metals?

Meet some soft reactive metals

Lithium, sodium and potassium are a group of elements on the far left of the periodic table. These metals do not behave as you might expect. They conduct heat and electricity very well, but they have some very different properties:

○ They are very reactive – you cannot leave them out without them reacting.

○ They become dull and tarnish very quickly in air.

○ They are not hard or tough – it is easy to cut them with a knife.

These pieces of lithium, sodium and potassium have just been cut with a sharp knife. They are so reactive that the freshly cut surfaces are soon becoming dull. This is because the metals react with oxygen and moisture in the air.

lithium

sodium

potassium

Corrosion

Another name for metals reacting in the air is **corrosion**. When iron corrodes it forms red flaky iron oxide, which we know as **rust**. Like lithium, sodium and potassium, iron reacts with the oxygen and moisture in air. When metals corrode they usually form metal oxides or hydroxides.

Tarnishing describes the reaction of less reactive metals like copper or silver, which just discolour on their surface.

This ship is made of steel (which is mainly iron). It was useful for many years but now it is rusting away.

Copper tarnishes to this green colour, but it does not corrode away like iron.

Guess what?

Lithium metal has such a low density that it floats on water. But a ship built from lithium would not last long!

Are metals materials that are shiny?

Well glass shines too but it is not a metal. Most metals are shiny but some go dull in air and water.

1 Copy and complete using words from the Language bank:

Metals react with air in different ways. It is the _____ in air along with moisture that cause this _____ .

2 What does tarnishing mean?

3 What is formed when iron reacts with air and moisture?

4 Why is lithium unsuitable for ship-building?

5 Find out the answers to these questions.
   a How do we remove tarnishing from metals like silver?
   b How can we stop metals tarnishing?

Language bank

corrosion
dull
moisture
oxygen
reactive
rust
shiny
tarnished
unreactive
water

Metals and water

How do metals react with water?

A gold ring does not react with water, even if the water is very hot. But some metals do react with water. Here is a general equation:

metal + water → metal hydroxide + hydrogen

The metal replaces a hydrogen in the water. The metal hydroxide and hydrogen are formed.

The metal caesium reacts so quickly with water that it explodes, shattering the container. Caesium is an extremely reactive metal.

Looking at potassium, sodium and lithium with water

Potassium reacts so violently it bursts into flames, burning the hydrogen that is formed.

Sodium reacts less vigorously without the hydrogen burning, but it gives out so much heat that it melts.

Lithium fizzes gently and skates around on the surface of the water.

Testing the products

You can collect the gas and test for hydrogen using the squeaky pop test. The water in the trough is alkaline. The metal hydroxide makes it alkaline, as you can see in the diagram opposite

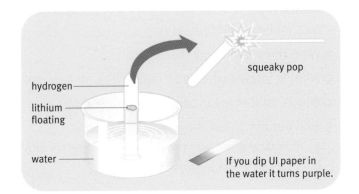

squeaky pop

hydrogen

lithium floating

water

If you dip UI paper in the water it turns purple.

Slightly slower

Calcium (left) is not as reactive as potassium, sodium, or lithium. It fizzes gently in water. Magnesium (right) reacts very slowly, showing a few bubbles on its surface after a couple of hours. It's faster using hot water.

Copper and unreactive metals like silver do not react at all with water. This is why we can use them in contact with watery liquids.

Back to the reactivity series

On page 59 we listed metals in order of how quickly they react with acids. We can also make a reactivity series based on how quickly metals react with water, as shown on the right.

The two lists are the same. This is because when a metal reacts with acid or water, it takes the place of hydrogen.

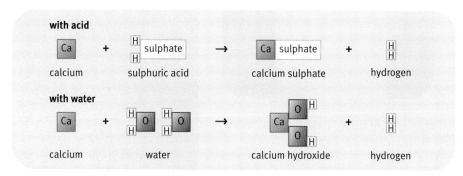

This simplified diagram shows how calcium takes the place of hydrogen.

Reactivity series

potassium (K)

sodium (Na)

calcium (Ca)

magnesium (Mg)

aluminium (Al)

zinc (Zn)

iron (Fe)

lead (Pb)

(hydrogen (H))

copper (Cu)

silver (Ag)

gold (Au)

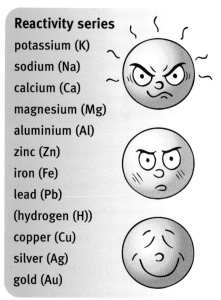

1 Copy and complete using words from the Language bank:

Some metals react with _____ to form the metal hydroxide and _____ gas. Other metals do not. We can write a _____ by comparing how quickly metals react with water.

2 Give two pieces of evidence to show that a chemical reaction takes place between sodium and water.

3 What is a reactivity series?

4 A label on a bottle of sodium recommends that the metal is kept under oil. Explain why this is.

5 Write a word or symbol equation for the reaction of calcium with water. (Hint: The formula for calcium hydroxide is $Ca(OH)_2$.)

Language bank

alkaline

hydrogen

metal

metal hydroxide

reactive

reactivity series

unreactive

water

Metals and oxygen

○ **Can we make predictions about the reactions of metals with oxygen?**

You have probably seen magnesium burning to make magnesium oxide. The metal reacts with oxygen as it burns. You can burn it in air, because there is oxygen in air, or you can fill a gas jar with oxygen from a cylinder and burn it in pure oxygen.

When something reacts with oxygen an oxide is formed. The reaction is called **oxidation**.

Magnesium burns well in air (left) to form magnesium oxide. It will burn with an even brighter flame in pure oxygen (right). Warning: do not look directly at burning magnesium as it can hurt your eyes.

Here is a word equation for the reaction:

magnesium + oxygen → magnesium oxide

A general equation for the reaction of a metal with oxygen is:

metal + oxygen → metal oxide

You can tell a reaction is happening because:

○ a new material is formed that is different from the metal
○ heat and light energy are given out – energy is transferred in chemical reactions.

If you weighed the metal before and after the reaction, you would find that its mass increased as it reacted with oxygen.

Magnesium oxide contains magnesium and oxygen atoms.

Iron reacts with oxygen to form iron oxide. This is the reaction that happens in sparklers.

Remember the reactivity series

Page 71 lists the metals in order in the reactivity series. We can use this to predict how each metal will react with oxygen. We would expect potassium, sodium and lithium to react more violently than magnesium, and copper and silver to react slowly.

Even a tiny amount of sodium burns violently in oxygen. Sodium oxide is formed.

Copper burns with a green flame to form copper oxide. It reacts much more slowly than sodium.

More about reactivity

You can think of reactivity as a drive to react with other things. Generally chemicals 'want' to be as stable as possible. Gold is already stable as it is. For others such as potassium, they are more stable if they react and combine with other substances.

More about oxides

If you put a metal oxide such as sodium oxide into water, the solution is alkaline. You can see this if you test the pH using universal indicator. On the other hand, solutions of non-metal oxides are usually acidic.

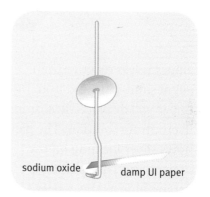

sodium oxide   damp UI paper

Burning sodium produces sodium oxide. If you touch this with damp UI paper, you can see that it forms an alkaline solution.

1 Copy and complete using words from the Language bank:

Metals react with oxygen to form a metal _____ . Some metals burn brightly while others react more slowly, depending on their _____ . A reaction with oxygen to form an oxide is called an _____ reaction.

2 Write each beginning with its correct ending.
Metal oxides tend to be …   … acidic.
Non-metal oxides tend to be …   … basic.

3 The order of reactivity of metals with acids and with oxygen is the same. Give examples that show this.

4 Write word equations for the oxidation reactions shown on these two pages.

Language bank ○—

acidic
alkaline
oxidation
oxide
oxygen
pH
reactivity

Metals pushing in

(M)

○ **Can metals displace each other?**

It's a knock-out

You have seen how we list metals in order of reactivity depending on how quickly they react with acid, water or oxygen. There is another way of comparing the reactivity of different metals.

If you have a compound of a metal, then a more reactive metal will knock out or **displace** this metal from its compound. You can see this if you have a solution of a metal salt and add a more reactive metal to it. The picture shows an example, adding magnesium to copper sulphate.

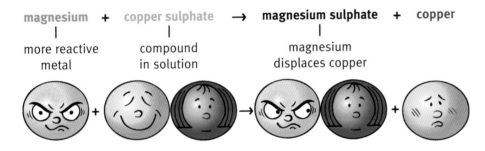

magnesium   +   copper sulphate   →   **magnesium sulphate**   +   copper

| | | |
more reactive metal   compound in solution   magnesium displaces copper

Magnesium displaces copper from copper sulphate. Magnesium sulphate solution is formed, which is colourless. Copper is deposited on the magnesium and falls to the bottom of the tube.

Copper is less reactive than magnesium, zinc and iron. But copper will displace silver from silver nitrate solution. The picture shows silver forming on the copper wire. The solution is turning blue as the copper salt is formed.

Copper is more reactive than silver, so copper displaces silver from silver nitrate solution.

Salt solution / Metal added	magnesium sulphate	zinc sulphate	iron sulphate	copper sulphate
magnesium		zinc deposited	iron deposited, green solution loses its colour	copper deposited, blue solution loses its colour
zinc	no reaction		iron deposited, green solution loses its colour	copper deposited, blue solution loses its colour
iron	no reaction	no reaction		copper deposited, blue solution loses its colour
copper	no reaction	no reaction	no reaction	

Some displacement reactions

The table shows what happens if you add some metals to solutions of other metal sulphates. The more reactive metal displaces the less reactive metal from its compound. If there is no reaction, this shows that the metal you are adding is less reactive than the metal in the sulphate.

A model for displacement

We can explain what is going on in displacement reactions by comparing them with something else. We call this sort of explanation a **model** or an **analogy**.

This is what is happening.

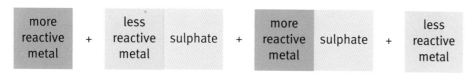

Think of reactive metals as being able to hold onto things they are joined to better than less reactive metals can. Magnesium holds onto the sulphate better than copper, so it displaces the copper in the compound. This leaves the copper all on its own.

The pull of the reactive magnesium is greater than the pull of the copper so magnesium wins and displaces copper.

1 Copy and complete using words from the Language bank:

   A metal will _____ a less _____ metal from its compound. It is as if the more _____ metal has a greater pull so holds onto the compound better.

2 What is an analogy?

3 Imagine that you have a friend who has missed a lesson on displacement reactions. Explain to your friend what you know so far about displacement reactions.

4 Choose three reactions shown on these two pages and write word or symbol equations for them.

Language bank

analogy
compound
displace
displacement
model
reactive
reactivity
replace
thermit reaction

Extracting and using metals

○ How does the activity series relate to uses and sources of metals?

The best metal for the job

Have you ever thought why knives and forks are made of steel (iron), or silver? Can you imagine what it would be like if they were made of a metal like sodium, which reacts violently with water? Life would certainly be interesting. And why do people have gold teeth rather than magnesium teeth?

Different metals have their own physical and chemical properties which make them good for certain jobs. These properties depend on their reactivity. Sodium is too reactive for everyday use, but gold is a very unreactive metal and does not react with saliva or food.

Aluminium or steel are good materials for car bodies and engine parts because the properties of the metals are well suited to the job. They can be shaped easily, they are strong and they are not too reactive.

Extracting metals

How we find a metal in the ground, and the method we use to extract it from its ore, is linked to its reactivity.

Unreactive metals like gold and silver are found in the ground **native** (as the metal element rather than reacted in a compound). But reactive metals quickly combine with materials in the ground to form compounds which we call **ores**.

> **Remember**
> A **property** describes how the material behaves.
> • Physical properties are things like hardness, strength, density and ductility.
> • Chemical properties are how it reacts, such as its reactivity with oxygen, acids and water.

If you were very lucky you might find pure gold in the ground or on a river bed.

This ore contains iron oxide. We need a reaction to extract the iron.

o We don't need a chemical reaction to extract gold or silver.

o To extract more reactive metals like zinc or iron we heat the ore with carbon. The carbon displaces the metal.

o For very reactive metals that are more reactive than carbon, we need to pass electricity through the ore to extract the metal. The process is called **electrolysis**. Electrical energy is used to split the melted or dissolved ore and separate out the metal. It is a very expensive process.

Could hydrogen gas be used to displace copper from copper oxide?

Yes, but it would be expensive. Carbon is cheap so we use that instead.

Metal	Extracted by	
potassium	electrolysis	
sodium	electrolysis	
calcium	electrolysis	more reactive than carbon so carbon can't be used to extract these metals
magnesium	electrolysis	
aluminium	electrolysis	
(carbon)		
zinc	smelting with carbon	
iron	smelting with carbon	less reactive than carbon so carbon will displace these metals from their ores
lead	smelting with carbon	
(hydrogen)		
copper	sometimes found native, sometimes as an ore which can be extracted with carbon	less reactive than hydrogen so these metals do not react with dilute acids or water
silver	found native	
gold	found native	

Guess what?

Aluminium is a very useful metal but it is difficult to extract. People discovered it much later than iron and copper. This is why we see iron, copper, silver and gold finds but no aluminium objects in ancient archaeological sites.

1 Copy and complete using words from the Language bank:

Very reactive metals are held in their _____ so strongly that we need to use electrical energy to separate them out. We call this process _____ . Less reactive metals can be extracted by heating with carbon which _____ the metal from its ore. This process is called _____ . Very unreactive metals are found _____ in the ground and rivers.

2 Aluminium is quite a reactive metal.
  a Is it easy or hard to extract from its ore?
  b Is it more or less reactive than iron?
  c Was it first used before or after iron?

3 Tin comes between iron and lead in the reactivity series. Predict what method is used to extract tin from its ore.

4 Find out more about the connection between the reactivity of a metal, when it was first used and how it is extracted from its ore.

5 Find out what bronze is. What is it used for? When and what was the Bronze Age? Was it before or after the Iron Age?

Language bank

chemical property
compounds
displaces
electrolysis
extract
extraction
native
ores
physical property
property
reactivity series
smelting

Checkpoint

1 Metals reacting

Midge reacted five metals with water. She listed them in order, fastest reaction first:

> potassium
> sodium
> calcium

Her teacher wouldn't let her try caesium with water as she said it was too dangerous.

Midge also reacted five metals with dilute hydrochloric acid. Here is her list:

> calcium
> magnesium
> zinc

Her friend told her that he'd tried copper with dilute hydrocloric acid and nothing happened. She knows that copper tarnishes in air, while gold stays shiny for years and years.

Copy and complete the following sentences, choosing the correct words.

a A list like the ones Midge made is called an **action list / a reactivity series / a reactivity cycle**.

b If Midge reacted the metals with oxygen, she would expect zinc to react **more quickly than / about the same as / slower than** sodium.

c If Midge combined her lists into one, she should add gold at the **top / bottom** of the list while caesium should be at the **top / bottom**.

2 Writing equations

Write word and symbol equations for the following metals reacting with oxygen.

a magnesium

b copper

c calcium

d iron

3 In order

Predict which reactions in question 2 will be fastest and slowest. List the four metals in order, with the one that reacts fastest with oxygen first. (You can use the information in question 1 to help you.)

4 About metals pushing in

Copy and complete these sentences, unscrambling the words.

A more reactive metal will replace a less reactive metal from its **poodmunc**. This is called a **nicetemsplad** reaction. If you add magnesium to copper sulphate solution, the **gemumians** displaces the copper. Magnesium **peshutla** and copper metal are formed.

5 Will it displace?

Predict whether a displacement reaction will happen if you mix these together. If yes, write a word equation. If you can, try to write a symbol equation for the reaction too.

a zinc and copper sulphate solution

b iron and sodium chloride solution

c calcium and zinc nitrate solution

d gold and copper nitrate solution

e magnesium and iron sulphate solution

Environmental chemistry

Before starting this unit, you should already be familiar with these ideas from earlier work.

○ Soil is a mixture of weathered rock particles, humus, water and air. What makes up humus?

○ Soils in different places may have different properties depending on the kind of rock they were formed from.

○ We can find the pH of a solution using universal indicator, or a pH meter. A solution that turns universal indicator red is strongly acidic, with a pH of 1. What pH would a neutral solution have?

You will meet these key ideas as you work through this unit. Have a quick look now, and at the end of the unit read them through slowly.

○ Some human activities, such as burning fossil fuels to generate electricity, pump polluting gases out into the atmosphere. Some of these gases react with rainwater to form an acidic solution called **acid rain**.

○ Acid rain takes part in neutralisation reactions with basic substances in rocks and building materials, causing damaging chemical weathering.

○ Acid rain also affects lakes, rivers and the soil, damaging plants and animals.

○ Scientists monitor the air and water for pollutants so that we can try and control the amounts that we put into the air. We can take steps to control acid rain such as improving efficiency so that we burn less fuels, using catalytic converters on vehicles and removing acidic pollutants from the gases given out by power stations.

○ Carbon dioxide and other gases in the atmosphere act as **greenhouse gases**, helping keep the Earth warm. People are producing more and more carbon dioxide. The resulting increased greenhouse effect may lead to **global warming**, causing devastating changes in the world climate.

○ In the atmosphere is a layer of **ozone** gas, which absorbs harmful ultraviolet radiation from sunlight. Pollutants have damaged this ozone layer, forming a 'hole' where it is thin. This lets more ultraviolet radiation through, increasing our risk of skin cancer.

○ How are soils different from each other?

What is soil?

Soil contains:

○ weathered rock particles (**sand** and **clay**)

○ **humus** (dead and decayed plant and animal material)

○ water

○ air

○ living material, including plant roots and animals such as worms and beetles, which change the soil as they live and grow there.

soil
↑
smaller weathered fragments
↑
weathered rock fragments
↑
solid rock

Even plants growing indoors need soil to anchor their roots and provide water and nutrients.

Soil supplies plants with valuable nutrients such as nitrates. It also provides them with water and somewhere to anchor their roots.

The properties of soil vary depending on the mix of rock particles. If there are lots of sand particles, water will drain through quickly. Clay particles are much smaller than sand. If there are lots of clay particles, the soil will be heavy and may become waterlogged. Lots of humus improves the properties of all soils.

Different soils, different places

Soils in different areas have different pHs. A soil may be acidic, alkaline or neutral. The pH of a soil depends on:

○ The minerals in the weathered rock particles that make up the soil. This varies according to the rocks in the area.

○ The plants growing there. Some plants add acidic material to the soil, and different plants absorb their own mix of nutrients from it.

○ The amount of rain that falls on the soil. Rainwater contains dissolved carbon dioxide, which makes it weakly acidic.

Guess what?

Rocks contain some minerals that act as a base. These react with acid, neutralising it. The neutralisation reaction weakens the rock.

o The amount of pollution in the area. Fossil fuels are burned in vehicles and in power stations, producing acidic gas pollutants. These dissolve in the rainwater and cause **acid rain**, which lowers the pH of the soil it falls on.

Cranberries and soil chemistry

Cranberries are an increasingly popular fruit. Cranberry juice is tasty and high in vitamin C. It is claimed to have many health benefits, such as preventing heart disease and cancer, and treating some infections.

You can check the pH of soil using a meter or soil test kit.

add lime to raise the pH

add gypsum to lower the pH

soil too acidic for crop

soil too alkaline for crop

You can change the pH of your soil using chemicals.

But if you want to grow cranberries, you'd better check your soil pH first. Cranberries grow well in acidic soils, with a pH between 4.2 and 5.5. If your soil pH is 7, cranberries will not grow well for you. You could grow peaches instead, as they prefer a soil pH of 6.5–7.

Another approach is to add chemicals to change the soil pH. Adding gypsum (magnesium sulphate) lowers soil pH. Adding lime (calcium oxide) raises soil pH.

1 Copy and complete using words from the Language bank:

Soil contains _____ and clay particles which formed from weathered rock. It also contains _____, which is the remains of dead animals and plants. Its _____ depends on the type of rock, the plants that grow there and the amount of _____ which falls.

2 What does soil provide for a plant?

3 Find out how you would test the pH of a sample of soil.

Language bank

acid rain
acidic
alkaline
clay
gypsum
humus
lime
pH
sand
soil

81

○ What happens to rocks and building materials over time?

Standing the test of time

These two monuments are in two very different environments. The climate in London is much wetter and cooler than in Egypt. Both cities suffer pollution from cars and industry. On Cleopatra's needle you can see that more detail has been eaten away. The monument in Egypt has not been exposed to so much polution and acidic rain, or to freezing temperatures.

How do rocks and building materials change?

Rocks are made from individual crystals or grains of minerals. These are cemented together into lumps. Weathering weakens rocks, breaking them up into smaller pieces and eventually into individual grains and crystals. The diagram on the opposite page reminds you how rocks and buildings become weathered.

Cleopatra's Needle dates back to about 1500 BC, and came to London in 1819.

This monument in Luxor, Egypt dates back to about 1300 BC.

If soil is acidic, the acid can weather rocks below the surface. Here the rock exposed to air is weathered very slowly, but below the surface of the soil chemical weathering has been much quicker.

Guess what?

How quickly a rock weathers depends on rock type, climate, soil type, local pollution levels, which plants are growing and even which way the rock faces.

Physical weathering

Exfoliation: the surface of the rock is heated through the day, when it expands. It is cooled at night, when it contracts. This regular expansion and contraction of the rock surface makes it flake off like peeling an onion.

Chemical weathering

Rainwater is naturally acidic because it contains dissolved carbon dioxide. It reacts with carbonate minerals in rocks, as shown by the equations below. Acid rain, which contains more strongly acidic pollutants, is even more damaging. It reacts more quickly with minerals in the rock, causing it to fall apart.

Physical weathering

Freeze–thaw weathering: water is an unusual chemical in that it expands when it solidifies. Liquid water seeps into small cracks. When the temperature falls, the water freezes and expands, making the cracks bigger. Eventually this can cause lumps of rock to fall off.

Biological weathering

Plant roots can grow into cracks in rocks. As they grow they can push the rock apart, making the crack bigger. Eventually bits of the rock start to fall away, weathering the rock. Small burrowing animals living in cracks can also weather the rock.

The reactions of chemical weathering by rainwater

carbon dioxide + water → carbonic acid (weak acid)

calcium carbonate + carbonic acid → calcium hydrogencarbonate (soluble)

Plants cause biological weathering to both rocks and buildings.

Language bank

biological weathering
carbon dioxide
carbonic acid
cemented
chemical weathering
climate
crystals
exfoliation
freeze–thaw weathering
grains
minerals
physical weathering
rainwater
weathering

1 Copy and complete using words from the Language bank:

Rocks and the buildings made from them contain small crystals and _____ of minerals, _____ together in lumps. These rocks are worn away by the processes of _____. There are three main types of weathering: physical weathering, _____ and _____.

2 **a** Describe two forms of physical weathering.

b What sort of climate will speed up physical weathering?

3 How can acidic soils weather rocks?

4 What do you think is the most important factor that has made Cleopatra's needle weather more quickly since it came to London? Explain which processes have caused this.

What causes acid rain?

You know that rocks are chemically weathered by rainwater, because it is naturally slightly acidic. Carbon dioxide in the atmosphere dissolves in rainwater, forming the weak acid carbonic acid. Sometimes rain can have an even lower pH. This **acid rain** speeds up the decay of building materials and paints, making rocks and buildings weather more quickly. It can make rivers, streams and lakes more acidic, killing plants and fish. It affects the soil so that trees and other plants cannot grow in it. Acid rain is a serious environmental problem.

How does acid rain form?

The burning of fossil fuels produces carbon dioxide and water vapour. It also produces other gases such as sulphur dioxide and nitrogen dioxide. Like carbon dioxide these also dissolve in rainwater, but they form stronger acids so are more corrosive. The acidic pollutants may be carried many miles by the wind before they fall as acid rain.

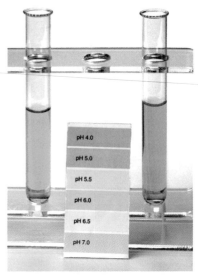

Carbon dioxide is a non-metal oxide, and these tend to be acidic. Rainwater containing dissolved carbon dioxide has a pH of about 6, but more acidic pollutants reduce its pH even further.

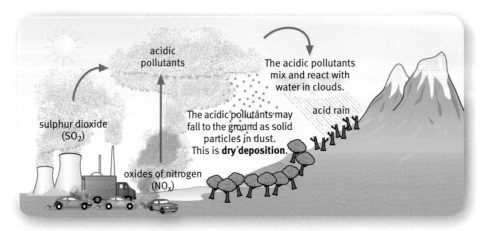

Sulphur dioxide and nitrogen oxides are non-metal oxides. They react with rainwater to make strong acids, causing acid rain.

Where do the acidic pollutants come from?

Fossil fuels such as petrol, diesel, coal and oil contain **hydrocarbons**. When these are burned the carbon and hydrogen are oxidised.

carbon + oxygen → carbon dioxide
hydrogen + oxygen → hydrogen oxide (water)

Fossil fuels also have impurities that contain other elements apart from carbon and hydrogen. These include sulphur and nitrogen, which are also oxidised when the fuel burns.

sulphur + oxygen → sulphur dioxide
nitrogen + oxygen → oxides of nitrogen

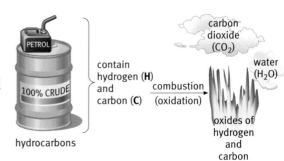

These oxides are acidic pollutants. With water in clouds they form acids such as sulphuric and nitric acids, which make acid rain.

Nitrogen oxides come mainly from vehicle engines, along with other industrial processes. We can't usually see acidic pollutants, but nitrogen dioxide can form brown smog.

Burning coal in power stations is a major source of sulphur dioxide. This pollutant also comes from smelting metal ores, and industrial burning and incineration processes.

Guess what?

Nitrogen forms many different oxides, including:
- nitrogen oxide (NO)
- nitrogen dioxide (NO_2)
- dinitrogen oxide (N_2O).

Scientists use NO_x to represent the mixture of oxides of nitrogen that form. Sometimes SO_x is also used to represent a mixture of different oxides of sulphur.

Where do acidic pollutants go?

Acidic pollutants can be carried long distances in the atmosphere before falling as acid rain. For example, the prevailing winds blow from the UK to Scandinavia, so that is where our pollutants tend to go. In the 1980s Britain was known in Scandinavia as 'the dirty old man of Europe'. Emissions have been reduced but the UK still produces pollution.

Acid rain causes tree leaves to go yellow and fall off, killing the trees. Some parts of Germany also suffer from the effects of acid rain.

Sulphur dioxide and nitrogen oxides are also produced by natural processes including volcanoes and lightning.

Language bank

acid rain
acidic pollutants
carbon dioxide
chemical weathering
dry deposition
fossil fuels
hydrocarbons
nitric acid
nitrogen oxides, NO_x
non-metal oxides
sulphur dioxide, SO_2
sulphuric acid

1 Copy and complete using words from the Language bank:

_____ makes rainwater slightly acidic. _____ is more strongly acidic, and this is caused by acidic pollutants dissolving in rainwater. These pollutants include _____ and _____.

2 Which human processes are the main producers of:
   a sulphur dioxide   **b** nitrogen oxides?

3 Draw a flow diagram with equations to explain how acid rain forms. Include the following words, and add a diagram if you want:

   sulphur dioxide   nitrogen oxides   pollutants   dissolve   acidic

The effects of acid rain

○ What are the effects of acid rain and how can they be reduced?
○ Is pollution worse now?

What does acid rain do to limestone rocks?

Chalk and limestone are mainly calcium carbonate. They react with acids, for example:

calcium carbonate + sulphuric acid → calcium sulphate + carbon dioxide + water

calcium carbonate + nitric acid → calcium nitrate + carbon dioxide + water

King's College Chapel in Cambridge has been standing for about 500 years. It is made of limestone so it is being weathered by acid rain. This has been happening for many years, and is still going on today.

Guess what?

The salt calcium sulphate is insoluble, so if it forms on the surface of a rock it can actually slow down any further weathering.

What does acid rain do to metals?

Metals such as zinc, iron and lead are used in buildings and sculptures. Exposed metals can react with acidic pollutants to produce a salt and hydrogen, for example:

zinc + sulphuric acid → zinc sulphate + hydrogen

Lead is less reactive than zinc and iron, so it is less affected by acid rain. Lead is used on roofs as it is soft and can be bent into shape.

Lead is unreactive enough to be used for roofing.

What does acid rain do to living things?

Plants can tolerate a range of pH, but very acidic rain reacts with minerals in the soil. This allows some useful minerals to be washed away, and can also cause toxic aluminium to be released into the soil. These changes to the soil damage the plants growing there.

pH	4.0	4.5	5.0	5.5	6.0	6.5
trout						
perch						
frogs						
snails						

Key ▓ danger ░ acceptable pH

A change in pH can seriously affect the food chains in a habitat.

Aquatic animals are sensitive to pH changes, especially when young. At pH 5 most fish eggs cannot hatch. Most lakes and rivers have a pH between 6 and 8, and if they are on carbonate rock this will help neutralise acid rain. But some lakes and rivers become so acidic that nothing can survive. Toxic aluminium is washed into the water from the surrounding soil, and this kills fish and insects. Lakes that contain no life are crystal clear.

Reducing acidic emissions

We have reduced our emissions of sulphur and nitrogen oxides over the last few years by the following:

o Using less fossil fuels by making engines more efficient, switching things off when we are not using them, or using alternative energy resources. This also helps us conserve supplies of fossil fuels.
o Using **catalytic converters**. Cars with petrol engines are fitted with devices which convert carbon monoxide (a poisonous gas) to carbon dioxide, and also convert oxides of nitrogen to nitrogen.
o Using **scrubbers**. These are devices in the chimneys of power stations and factories. They spray the waste gases with a scrubbing agent, which reduces sulphur dioxide emissions.

These measures have all helped, but many lakes affected by acid rain still have not recovered.

*Some kinds of lichen will only grow in clean air. Scientists can use them as **indicator organisms**. The types of lichen growing in an area give an indication of how clean the air is.*

Language bank

acid rain
aluminium
carbonate
catalytic converters
cement
chalk
indicator organisms
iron
lichens
limestone
minerals
reactive
scrubbers
sulphuric acid
zinc

1 Copy and complete using words from the Language bank:

Acid rain reacts with rocks and buildings made of _____. It also reacts with exposed metals such as zinc and _____. Acid rain affects the _____ in soil and damages both plant and animal life.

2 What measures can we take to reduce air pollution?

3 a Look at the diagram above. Which animal is least tolerant of acidic conditions? Which is most tolerant?
b How might acidic rain pollution affect organisms in a pond?

Global warming: good or bad?

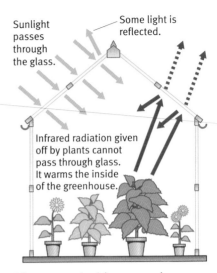

Sunlight passes through the glass.

Some light is reflected.

Infrared radiation given off by plants cannot pass through glass. It warms the inside of the greenhouse.

It's warmer inside a greenhouse.

○ Is global warming happening?

Plants grow better in a greenhouse because it's warmer inside than outside. Sunlight passes through the glass. The plants and other objects inside the greenhouse absorb the light, and they give out some infrared radiation. This cannot pass through the glass, so it is kept inside the greenhouse, warming the air and everything else inside.

The atmosphere around the Earth acts like the glass in a greenhouse. Sunlight passes through the atmosphere to the Earth's surface. The Earth gives out infrared radiation, and some gases in the atmosphere reflect this back to the Earth rather than letting it go out into space. This keeps the Earth warmer than it would otherwise be. Without this **greenhouse effect**, life as we know it would not exist.

So the atmosphere keeps us warm.

Yes, like a blanket around the Earth.

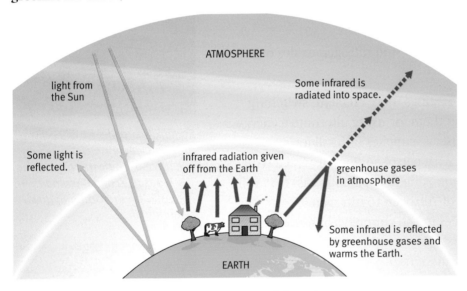

ATMOSPHERE

light from the Sun

Some infrared is radiated into space.

Some light is reflected.

infrared radiation given off from the Earth

greenhouse gases in atmosphere

Some infrared is reflected by greenhouse gases and warms the Earth.

EARTH

Greenhouse gases: what's the problem?

The gases in the atmosphere that reflect the infrared radiation are called **greenhouse gases**. They are shown in the table.

Guess what?

If we had no atmosphere, Earth temperatures would be about 33 °C lower.

Greenhouse gas	Produced by
carbon dioxide	burning fossil fuels for generating electricity and for transport; also by respiration, rotting organic material, volcanic eruptions
methane	rotting organic material (agriculture and landfill sites)
HFCs (hydrofluorocarbons) CFCs (chlorofluorocarbons)	CFCs used to be used in aerosols and fridges but they caused damage to the ozone layer (see page 90). Now HFCs are used instead.
NO$_x$	traffic pollution, also produced naturally by bacteria
water vapour	many natural sources, burning fossil fuels

The problem is that we are producing more and more greenhouse gases. Scientists think that because of this the greenhouse effect is increasing, making the Earth warmer. This is called **global warming**.

The Earth's average temperature has warmed by 0.6 °C over the last 100 years or so. Some people predict that the average UK temperature may rise by 2–3.5 °C by 2080. This does not mean we will all have better weather, as global warming can cause many climate changes:

o **Temperature change:** some areas may get warmer, others colder.
o **Flooding:** oceans expand as they get warmer, which may cause flooding. Millions of people's homes could flood regularly. The sea level may rise by 50 cm by 2080.
o **Rainfall changes:** some areas might become wetter, such as North America and South East Asia. Some areas might become drier, such as Africa and India.
o **Plant life:** some areas might be able to grow better or different crops. Others may not be able to grow crops at all.

Is our increased production of carbon dioxide really causing global warming? Look at the graphs. The first shows that we are putting more and more carbon dioxide into the atmosphere. The second graph shows how the average temperature is changing. There have always been fluctuations in temperatures, and it is difficult to be sure what the long-term trend will be. Not all scientists agree.

How can we reduce greenhouse gas emissions?
Use less fossil fuels:
o use alternative energy resources
o switch off electrical appliances when not in use
o use energy-efficient devices.
We should also stop deforestation (stop cutting down trees) and plant more trees.

1 Copy and complete using words from the Language bank:

There are gases in the _____ that act like a greenhouse. They help reflect _____ back to the Earth, warming it up. These gases are called _____ and they include _____ and methane.

2 **a** Explain how the atmosphere keeps the Earth warm.
  b Why is this effect increasing?

3 How would planting more trees help reduce global warming?

4 Compare the measures for reducing greenhouse gas emissions with those for reducing acidic pollutants on page 87.

5 Using a search engine, type in 'global warming' and try to find evidence to support the theory that global warming is happening.

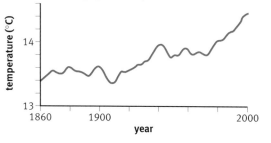

Carbon dioxide in the atmosphere since 1860

Compare these two graphs. Do you think carbon dioxide emissions are making the Earth warmer?

Scientists monitor lakes, rivers and the air to see if conditions are changing over time.

Language bank

atmosphere
carbon dioxide
fossil fuels
global warming
greenhouse effect
greenhouse gases
infrared radiation
methane

89

A problem with the ozone layer Ⓢ

What is ozone?

Ozone is a form of oxygen which is formed in the atmosphere. There is a thin layer of ozone in the atmosphere around the Earth, about 30 km up. This **ozone layer** absorbs ultraviolet (UV) radiation from the Sun, stopping much of it reaching the Earth's surface.

UV radiation can be harmful:

o It can cause skin cancer.

o It can make skin tight and non-elastic.

o It can damage your eyes.

Our natural defence against harmful UV radiation is a pigment in our skin called **melanin**. This is what makes our skin darker if we spend time in the sun. The dark-coloured melanin absorbs the harmful radiation, stopping it getting through the skin to the tissues beneath.

The hole in the ozone layer

The problem is that we are producing gases which react with the ozone layer, damaging it and letting more harmful UV radiation through. Chemicals called CFCs (chlorofluorocarbons) used to be used in fridges, aerosol sprays and air-conditioning units. These CFCs reacted with the ozone molecules, creating a 'hole' in the ozone layer. The hole allows more UV light to reach the Earth, increasing the risk of skin cancer and also having other effects such as killing sea plankton. Nowadays new chemicals called HFCs have largely replaced CFCs, and these do not react with ozone. However, some countries are still using ozone-destroying chemicals. It is hoped that all CFCs will be banned in Europe by 2015.

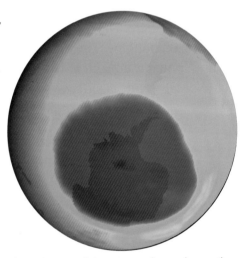

This picture of the atmosphere shows the hole in the ozone layer (dark blue) in September 2003. The 'hole' changes size at different times of year.

Confused by all this pollution?

Acid rain? The greenhouse effect? The hole in the ozone layer? It is easy to become confused about pollution. The table opposite summarises the three major problems that we face.

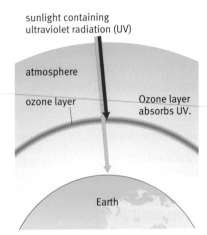

The ozone layer absorbs UV radiation.

> ### Guess what?
>
> Cars and fossil fuel power stations produce ozone gas. At the Earth's surface ozone is a pollutant, causing asthma and other lung problems. But the ozone layer high in the atmosphere protects us. We don't want ozone pollution at ground level, but we do want the ozone layer high in the atmosphere.

Problem	Pollutants that cause it
acid rain	sulphur dioxide and oxides of nitrogen
global warming	increased levels of greenhouse gases including carbon dioxide, methane, nitrogen oxides and CFCs
hole in the ozone layer	CFCs and other chlorine-based chemicals

Anti-pollution measures

Many countries have agreed to reduce the amount of carbon dioxide they put into the air each year. Between 1990 and 2000, sulphur dioxide emissions were reduced by 71%. The use of CFCs is declining and will soon be banned outright. So things are improving in some ways.

But every country wants industrial and economic development, to give its people a better standard of living. We must agree to develop in a way that does not destroy what we have. This is called **sustainable development**. Here are just a few views on how we can achieve this.

We should recycle waste paper, aluminium and glass instead of using up the Earth's resources to make new materials.

We should tax the sale of cars. This would help cover the cost of dismantling old cars in an environmentally friendly way.

We should tax fuels and electricity to fund research into alternative energy resources such as wave power which do not produce carbon dioxide.

We need to encourage people to car share and use public transport.

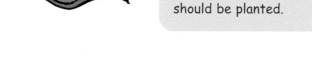

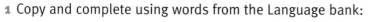

Trees absorb carbon dioxide and produce oxygen. When trees are cut down for paper or timber, more trees should be planted.

We need to monitor the environment to make sure we are not destroying valuable habitats for the plants and animals that live in our country.

1 Copy and complete using words from the Language bank:

   _____ is a form of oxygen which makes up a layer high in the atmosphere. This layer filters out harmful _____ which can cause skin cancer. Some chemicals including _____ have reacted with the ozone layer. We refer to this problem as the _____.

2 Summarise the differences between acid rain, global warming and the hole in the ozone layer. Start with the table opposite, and add two columns showing what activities cause the pollution and what problems it causes for us.

3 Look at the ideas above. For each one, explain how it will help in reducing pollution and sustaining the Earth's resources. Do you think it would be a measure that people would like to see introduced?

4 We all want to breathe clean air. Identify ten everyday things you and your friends could do to help save the Earth's resources and reduce pollution.

Language bank

acid rain
carbon dioxide
CFCs
global warming
greenhouse effect
hole in the ozone layer
ozone
recycling
sustainable development
UV radiation

Checkpoint

1 Matching reactions

Match up each environmental process with the correct chemical reaction.

Processes

Acid rain quickly weathers limestone rocks and buildings.

Naturally acidic rainwater slowly weathers limestone rocks and buildings.

Carbon dioxide dissolves in water.

Carbon dioxide is formed when coal and hydrocarbon fuels are burned.

Metals are corroded by acid rain.

Reactions

carbon + oxygen → carbon dioxide

carbon dioxide + water → carbonic acid

zinc + sulphuric acid → zinc sulphate + hydrogen

calcium + nitric → calcium + carbon + water
carbonate acid nitrate dioxide

calcium + carbonic → calcium
carbonate acid hydrogencarbonate

2 The greenhouse effect

Copy and complete these sentences, unscrambling the bold words.

Carbon dioxide and methane are **geeserhuno** gases. This means they help to reflect **raferdin** radiation back to Earth, keeping the Earth warm. This is effect called the **reshenogue tecfef**.

Humans are putting more and more greenhouse gases into the **pathosmere**. Scientists think this is causing **goball marwing**.

3 Indicator organisms

Look at the diagram on page 87. A scientist suspects acidic pollution in a lake, because during routine monitoring she has observed changes in the organisms that live there.

a If she has observed that perch and frogs are living there, but no trout or snails, what might you expect the pH to be?

b Which is the most acid-tolerant organism shown?

c Which mineral below is washed into water following acid rain pollution, killing fish?

nitrate   ammonium   aluminium

4 True or false?

Decide whether the following statements about the ozone layer are true. Write down the true ones. Correct the false ones before you write them down.

a Ozone is a form of the element hydrogen.

b The ozone layer absorbs harmful ultraviolet radiation.

c CFCs do not react with ozone.

d A hole has appeared in the ozone layer, exposing us to more ultraviolet radiation.

e This means more people are suffering from skin cancer and eye damage.

f The hole is getting bigger because everyone is still using CFCs.

5 Pollutants and problems

For each problem below, choose the pollutants and processes that cause it.

Problems

acid rain: dying trees and fish

global warming: climate change

hole in the ozone layer: increased skin cancer

Pollutants

increased carbon dioxide, methane, NO_x

CFCs

sulphur dioxide and oxides of nitrogen

Processes

used in fridges and air-conditioning units

fossil-fuel power stations and vehicles

increased combustion and cutting down forests

Using chemistry

Before starting this unit, you should already be familiar with these ideas from earlier work.

- In a chemical reaction, new materials are formed. List some signs that tell you a chemical reaction is happening.
- We describe chemical reactions using word equations or symbol equations.
- Some reactions give off a gas. We can test to see which gas it is.
- Burning is a chemical reaction. The burning material reacts with oxygen in the air. New compounds called **oxides** are formed.
- We can list metals in order according to how reactive they are. What do we call this list?

You will meet these key ideas as you work through this unit. Have a quick look now, and at the end of the unit read them through slowly.

- Many chemical reactions give out heat energy. A particularly useful one is the burning of a fuel, and we use this **oxidation** reaction to provide us with heat energy.
- Instead of giving out heat energy, we can sometimes arrange a chemical reaction so that it gives out electrical energy. This is the basis of the batteries that power many devices you use, and car batteries which allow drivers to start their cars.
- Some chemical reactions take in heat energy, and we can use these to cool things down.
- We use chemical reactions to make new products. Chemists are skilled at designing and making new materials, some synthetic and some natural. Plastics, drugs, foods, building materials and countless other products central to our lives depend on the work of chemists.
- In a chemical reaction, the total amount of material does not change. The mass of the products is the same as the mass of the reactants.
- The particle model explains this idea of **conservation of mass**. In a reaction, the particles are rearranged. No particles are added, and none are lost.
- Mass is also conserved when materials change state or dissolve.

Burning

○ **What chemical reactions take place when fuels burn?**

Burning fuels

A **fuel** is a substance that we burn to provide us with heat energy. This energy is what makes fuels valuable resources. Burning is also known as **combustion**.

Combustion is an **oxidation** reaction. The fuel reacts with oxygen to form oxides.

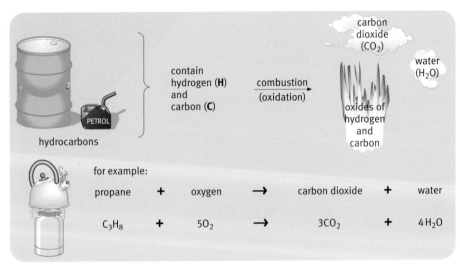

for example:

propane	+	oxygen	→	carbon dioxide	+	water
C_3H_8	+	$5O_2$	→	$3CO_2$	+	$4H_2O$

*Many fuels are **hydrocarbons** – they contain carbon and hydrogen. Hydrocarbons burn to produce carbon dioxide and water.*

Complete and incomplete combustion

When we burn a fuel, we want to release as much energy from it as we can. To do this the fuel needs to burn completely, forming just carbon dioxide and water. This is called **complete combustion**.

Sometimes there is not enough oxygen getting to the fuel as it burns, and then **incomplete combustion** may take place. This wastes the fuel, because not all the energy is released from it, and it also causes more pollution. The equations show what happens with coal, which is carbon.

carbon + oxygen → carbon dioxide, CO_2 **complete combustion**

carbon + oxygen → carbon monoxide, CO   **incomplete combustion**

As well as carbon monoxide, incomplete combustion can produce unburned carbon as soot.

What's the best fuel?

There are many fuels, and each has its own advantages and disadvantages. Look at the following table.

For lift-off the main engines burn hydrogen. Additional thrust comes from solid fuel in the white boosters.

Guess what?

Carbon monoxide is a dangerous, poisonous gas. It may be produced by the incomplete combustion of natural gas in badly maintained gas heaters. Carbon monoxide stops the blood carrying oxygen to the body cells. It often kills people while they are asleep.

Uni - Com
Carbon Monoxide Detector

Date Opened:	
CARBON MONOXIDE DETECTOR	DARK SPOT INDICATES DANGER

EMERGENCY CONTACT:

A simple carbon monoxide detector saves lives.

Fuel	State	Advantages	Disadvantages
coal	solid	gives out a lot of heat energy per kilogram of fuel	non-renewable fuel, sulphur impurities contribute to acid rain
		If you use coal you have to carry it inside, and clean out the ash.	
natural gas (methane)	gas	piped to your door, clean and convenient	not available everywhere, not sold in portable bottles
		Natural gas is very convenient, but it is not available everywhere.	
wood	solid	cleaner burning than coal	need to plant more trees to replace those used
		Wood needs to be cut and carried inside, but some people like a wood fire and trees grow again quite quickly.	
oil	liquid	useful for areas without natural gas	has to be delivered by tanker
		Heating oil can be delivered to homes where gas is not available.	
butane	gas	bottled, so very portable	has to be delivered by tanker

This helicopter needs to fill up with kerosene (paraffin) fuel every few hours of flying.

1 Copy and complete using words from the Language bank:

Burning is also called _____. A fuel is a substance that reacts with oxygen to form oxides and release _____. If oxygen is limited, fuels may undergo _____. This produces the poisonous gas _____.

2 a What fuel would be best to heat a top-floor flat in the middle of a city?

b On a remote Scottish island there is no natural gas supply. What might be the best fuel to use here?

c Explain why coal would be an unsuitable fuel for a helicopter.

3 In some ways hydrogen is an ideal fuel, because it burns to produce only water, giving out lots of energy. Use the Internet or textbooks to find out about the advantages and disadvantages of hydrogen as a fuel. Could it be the fuel of the future?

Language bank

carbon monoxide
combustion
complete combustion
energy
fuel
hydrocarbons
incomplete combustion
oxidation

Energy and reactions

○ **How else are chemical reactions used as energy resources?**

Energy resources

In a chemical reaction, the reactant particles are rearranged to form the products. Often the reactant particles are joined together. They have to be split apart, which takes energy, before they can reform to make the products, giving out energy.

In the combustion of a fuel, heat and light energy are released, so fuels are useful energy resources.

carbon atoms in coal   oxygen molecules   energy is needed to split up the molecules   lots of energy is given out when carbon dioxide molecules are formed

Exothermic reactions

A reaction such as combustion that gives out heat energy is called an **exothermic reaction**. On pages 74–5 you saw that in a displacement reaction, a more reactive metal displaces a less reactive metal from its compound. The thermit reaction is a displacement reaction which gives out lots of heat energy.

Striking a match is another exothermic reaction. But not all reactions release heat and light energy. Sometimes reactions release electrical energy, and some reactions take in energy rather than giving it out.

Electrical energy

We can use displacement reactions to give out electrical energy instead of heat energy. We put strips of different metals into a salt solution, to form a simple cell. The voltmeter shows that electrical energy is given out. The further apart the two metals are in the reactivity series, the higher the voltage they produce. The lead–acid batteries used in cars work on a similar principle, and you can see a lemon cell on page 109.

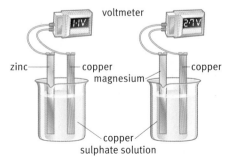

voltmeter

zinc — copper
magnesium — copper

copper
sulphate solution

Zinc displaces copper and a voltage of 1.1 V is produced. Magnesium is more reactive than zinc, and more electrical energy is given out.

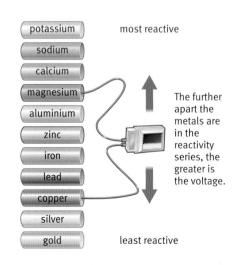

potassium — most reactive
sodium
calcium
magnesium
aluminium
zinc
iron
lead
copper
silver
gold — least reactive

The further apart the metals are in the reactivity series, the greater is the voltage.

HOW DO MATCHES WORK?

The Chinese were the first to use matches around 600 AD. Early match heads were just sulphur, which catches fire easily, but they often set themselves alight without being struck. In the mid-1800s an improved design called Lucifers or friction matches were used, which had phosphorus as well as sulphur. Today's safety matches are much safer.

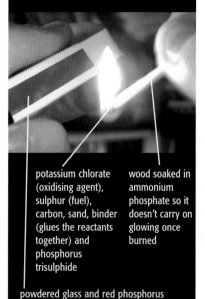

potassium chlorate (oxidising agent), sulphur (fuel), carbon, sand, binder (glues the reactants together) and phosphorus trisulphide

wood soaked in ammonium phosphate so it doesn't carry on glowing once burned

powdered glass and red phosphorus

Striking the match causes an exothermic reaction and starts the phosphorus burning in air. This ignites the sulphur and oxidising agent, which give a more controlled flame.

Endothermic reactions

Some reactions take in energy from the surroundings, rather than giving it out. We see this as a drop in temperature. These reactions are called **endothermic reactions**.

Reactions to warm you up . . .

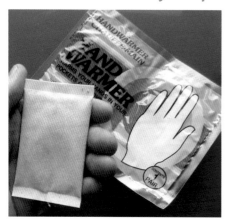

In these hand warmers, iron and water react to form iron oxide. It is an exothermic reaction, and the heat energy warms your hands. Other types of hand warmer contain melted chemicals. Clicking a disc triggers them to solidify, which is an exothermic process (but not a chemical reaction), so they give out heat.

The reaction between citric acid and sodium hydrogencarbonate solution is endothemic.

. . . and cool you down

If you have a sports injury, applying ice can reduce the bruising and swelling. Chemical cold packs are more convenient than carrying ice around. When the chemicals mix, an endothermic reaction happens. The cold pack cools the injured area.

The reaction in this cold pack takes in heat energy, cooling the injury and preventing swelling.

The reactions in fireworks give out light and sound energy as well as heat energy.

Guess what?

Sherbet contains solid citric acid (a weak acid) and sodium hydrogencarbonate. When you put this mixture in your mouth, they dissolve and react. The endothermic reaction gives off carbon dioxide, which gives sherbet its fizzy cool taste.

1 Copy and complete using words from the Language bank:

Some chemical reactions give out energy. These are called _____ reactions. Other reactions take in energy. These are called _____ reactions. Fuels store chemical energy which is given out as heat energy when they burn, so they are _____.

2 Look at the diagram of a cell on the opposite page, with two different metals in a salt solution. What can you say about the reactivity of the two metals and the voltage produced?

3 In camping shops you can buy coffee and soup with a self-heating device underneath the cup which heats the drink. Use the Internet or other sources to try and find out how these self-heating devices work.

Language bank

cell
displacement reaction
electrical energy
endothermic
energy resources
exothermic
heat energy
voltage

What types of new material are made through chemical reactions?

Chemistry is the study of matter. Chemists develop all kinds of materials from cosmetics to construction materials. They study what materials are made of, how they behave and how they can be used. They use chemical reactions to make or **synthesise** new substances.

In this lab, the chemists are synthesising new materials.

How new materials are made

All around you can see furniture, clothing, foods and equipment. **Synthetic** materials like plastics or new drugs have been made by reactions in factories. Natural materials such as wool or cotton have been made by reactions in living organisms.

There are several stages in the production of a new material.

	Stage	Scale	
refine and scale up	research: laboratory stage	bench chemistry (less than 1 kg)	
refine and scale up	development: pilot plant stage	larger scale (e.g. 1000 kg)	
	full production: factory	full scale (tonnes)	

During the development process, the new material and the process used to make it will be tested to make sure they are effective and safe. For example, drugs have to be tested on patients to make sure they work and do not have bad side effects. Plastics might be monitored for their strength. Research is also done to see whether people will want to buy the new material.

Materials science

All the substances you think of as 'plastics', such as polythene or PVC, are **polymers**. Polymers are very large molecules, built up from lots of repeated units called **monomers**.

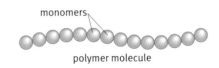

Polymers have different properties depending on their monomers.

Polymers are made from crude oil which is found in the Earth's crust. Oil companies drill down and pump the oil out of the ground or sea bed. Crude oil is a mixture, and it is separated at an oil refinery using **fractional distillation** as shown on the next page.

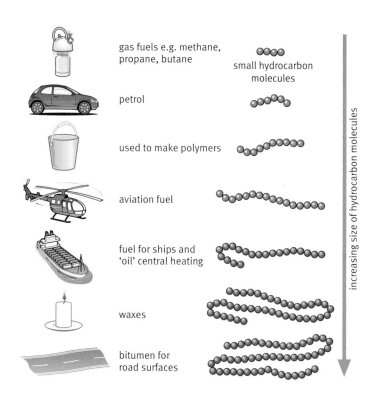

In fractional distillation, the crude oil is heated and different components boil off at different temperatures. The gases are collected and separated from each other.

Pharmaceuticals and food science

Chemists and biochemists work to produce new medicinal drugs every year. Some drugs are adapted from natural substances produced by plants or micro-organisms. Others are totally new synthetic molecules.

Chemists use living micro-organisms to produce foods such as yoghurt and cheese. Conditions of temperature and pH need to be carefully controlled for these living chemical reactions.

The chemical industry is huge and varied, and it employs a whole host of people apart from the scientists. Every factory will need accountants, engineers, builders, safety experts, sales staff and many more. From farming to forensics, chemistry touches our lives in many ways.

Special types of bacteria and fungi are used in controlled conditions to make each individual type of cheese.

1 Copy and complete using words from the Language bank:

Chemistry is the study of materials. Some materials are _____, which means they are made by people. An example is plastics, which we also call _____. Other materials are _____ – they are made by living things. The _____ industry uses chemistry to develop new drugs.

2 A Year 6 pupil wants to know more about chemistry. Explain to them what chemistry is, and its importance in our lives.

3 Sketch a few scenes to plan a short TV programme about the stages in the development of a new drug.

Language bank

chemistry
development
fractional distillation
materials
monomer
natural
pharmaceuticals
plastics
polymers
production
research
synthetic

Chemical reactions

○ What happens to atoms and molecules when new materials are made?

In a chemical reaction, the particles in the reactants are rearranged to form the products. No particles are lost, so the total mass stays the same.

The law of conservation of mass states that:

In a chemical reaction, mass is not lost or gained.

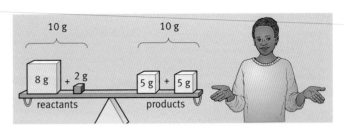

So if we have 10 g of reactants, we get 10 g of products.

Precipitation reactions

In a precipitation reaction, you mix two solutions and a solid is formed (a precipitate). Mass is conserved in a precipitation reaction.

If you mix silver nitrate and sodium chloride solutions, a white precipitate of silver chloride forms. Eventually this will fall to the bottom of the container. No material leaves the beaker, so the products weigh the same as the reactants, as the photos show.

silver nitrate + sodium chloride → silver chloride + sodium nitrate

Reactions that produce a gas

Some reactions give off a gas. If you put them on a balance, the mass goes down as the reaction proceeds. This is because the gas bubbles off and leaves the reaction container. This happens in the reaction between calcium carbonate and hydrochloric acid:

calcium carbonate + hydrochloric acid → calcium chloride + carbon dioxide + water

If you mix 20 g of calcium carbonate and 100 g of hydrochloric acid, you would expect them to make 20 + 100 = 120 g of products. But the carbon dioxide gas escapes from the flask, so the mass goes down.

Mass is conserved in a precipitation reaction.

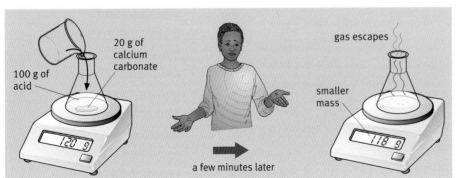

The mass goes down as the reaction proceeds, as gas is lost to the air.

To prove that mass is conserved, you can collect the gas and measure its mass. The mass of the products equals the mass of the reactants.

If you collect the gas, you can see that mass is conserved in the reaction.

Particles and conservation of mass

We can use the particle model to explain conservation of mass in chemical reactions. The particles in the reactants recombine to form the products. No particles are added or lost. For the precipitation reaction:

silver nitrate (soluble)   sodium chloride (soluble)   silver chloride (insoluble – precipitate)   sodium nitrate (soluble)

For the reaction between calcium carbonate and hydrochloric acid:

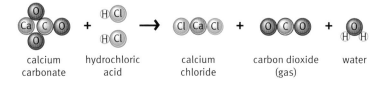

calcium carbonate   hydrochloric acid   calcium chloride   carbon dioxide (gas)   water

Guess what?

In some reactions there is just one reactant, which breaks down to form the products. Thermal decomposition is an example: if you heat calcium carbonate enough, it decomposes into calcium oxide and carbon dioxide.

Dissolving and changing state

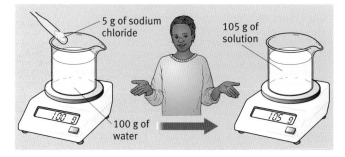

When you dissolve sodium chloride in water, the total mass is the same before and after.

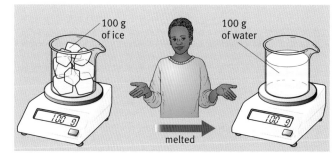

When you melt ice, the total mass is the same before and after. No particles are added or lost.

1 Copy and complete using words from the Language bank:

The law of _____ says that the mass does not change during a chemical reaction. No mass is lost or gained as the _____ turn into the products.

2 1.2 g of carbon burns and reacts with 3.2 g of oxygen. How much carbon dioxide is produced?

3 Why does melting 8 g of solid wax produce 8 g of melted wax? Use the word 'particles' in your answer.

4 Write symbol equations for the precipitation and thermal decomposition reactions described on these pages.

Language bank

changing state
conservation of mass
dissolving
particle model
precipitation
products
reactants

A closer look at chemical reactions

Magnesium reacts with oxygen to make magnesium oxide. There is an increase in mass as oxygen from the air combines with the magnesium to form magnesium oxide.

magnesium + oxygen → magnesium oxide

$$2Mg + O_2 \rightarrow 2MgO$$

How can we find out how much oxygen reacts with the magnesium?

Finding out how much magnesium and oxygen react together

1	Weigh a crucible with its lid.
2	Add a small piece of magnesium ribbon and weigh again. Note down these masses.
3	Put the crucible on a tripod with a pipe-clay triangle and heat strongly.
4	Occasionally lift the lid of the crucible to allow air in.
5	Allow the apparatus to cool. Reweigh the crucible, lid and contents.

Results		
mass of crucible and lid		=15.3 g
mass of crucible, lid and magnesium before heating		=20.1 g
so mass of magnesium	= 20.1 – 15.3 g	= 4.8 g
mass of crucible, lid and solid after heating		=23.3 g
so mass of magnesium oxide	= 23.3 – 15.3 g	=8.0 g
so mass of oxygen that reacted	=8.0 – 4.8 g	=3.2 g

You can see from the results that 4.8 g of magnesium produced 8.0 g of magnesium oxide. We can work out the mass of oxygen that reacted – it was 3.2 g.

Chemicals react in fixed amounts. Here 4.8 g of magnesium reacted with 3.2 g of oxygen to make 8.0 g of magnesium oxide. If we do the reaction with lots of different masses, we can plot the results on a graph. The graph is a straight line.

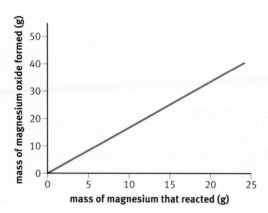

4.8 g   3.2 g   8.0 g

The magnesium atoms join up with the oxygen atoms to make magnesium oxide.

If you know the mass of reactant, you can work out the mass of product.

The total mass of reactants equals the mass of the product.

That's the law of conservation of mass, isn't it?

A little bit of burning history

You know that when things burn, such as metals or fuels, they combine with oxygen to form oxides. But in the eighteenth century people had another theory about what was happening. The German scientist Georg Stahl explained burning using the phlogiston model. This said that something (called phlogiston) comes out of a burning material, leaving behind the remaining material called calx. Even Joseph Priestly, who discovered oxygen, used this theory of burning. Eventually scientists noticed that things get heavier not lighter when they burn, as oxygen combines with the fuel in the combustion reaction. It was the Frenchman Antoine Lavoisier who made the connection between burning and oxygen.

According to the phlogiston theory, materials lose phlogiston (are dephlogisticated) when they burn!

1 Copy and complete using words from the Language bank:

When magnesium reacts with oxygen, magnesium oxide is formed. We can find the mass of oxygen that reacts with the magnesium using the law of _____.

2 Why does magnesium gain mass as it reacts with oxygen? Does this go against the law of conservation of mass?

3 Look at the graph. If you heated 8 g of magnesium, how much magnesium oxide would you expect to get?

4 Find out more and write a short paragraph about each of the following:

a phlogiston   **b** Priestley   **c** Lavoisier.

Language bank

conservation of mass
Lavoisier
phlogiston
Priestley
products
reactants

Checkpoint

1 Burning questions
Match up the beginnings and endings to make complete sentences.

Beginnings

The burning reaction

Heat energy is given out,

Fuels give out lots of

The fuel reacts with oxygen,

If there is not enough oxygen,

Endings

heat energy when they burn.

is also known as combustion.

so it is an exothermic reaction.

incomplete combustion may occur.

so it is an oxidation reaction.

2 Rearranging particles
Here is a particle diagram for burning coal. Sketch the diagram and choose the correct label below for A–D.

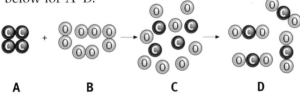

A   B   C   D

Labels

oxygen molecules

carbon dioxide forms: lots of energy out

carbon atoms

energy in

3 Electrical energy
Look at the following diagram of a reaction giving out electrical energy. Which pair of metals below would give the highest reading on the voltmeter?

○ zinc and iron

○ magnesium and silver

○ zinc and silver

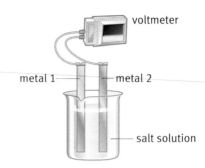

voltmeter

metal 1 — metal 2

salt solution

4 Made by chemistry
Draw a large concept map to show the importance of chemistry, and some uses of the many new materials that chemists make. Include these words, and add examples of products for each one:

synthetic   food

natural   fertilisers

polymers   crude oil

drugs   drinks

5 Conserving mass
In a chemical reaction, mass is conserved. The mass of the products equals the mass of the reactants. Choose the best statement below that explains why this is.

○ Energy may be given out in a chemical reaction.

○ There are no new particles in a chemical reaction – the reactant particles are just rearranged.

○ A reactant or product may dissolve and disappear.

○ When you heat magnesium in a crucible, you can predict how much magnesium oxide should be formed.

○ The mass will go down if a gas is given off.

Energy and electricity

Before starting this unit, you should already be familiar with these ideas from earlier work.

- A cell or battery is the energy source in many electrical circuits. It pushes current around the circuit.
- In the components, electrical energy is transformed. What energy transformation happens in a bulb?
- In a series circuit, there is only one route for the current. The current is the same all the way around the circuit. What happens if a bulb blows in a series circuit?
- In a parallel circuit, there is more than one route for the current. There may be a different current in each loop.
- We need energy to make things happen. We can burn fuels to provide energy.
- Renewable energy resources help us conserve fossil fuels.

You will meet these key ideas as you work through this unit. Have a quick look now, and at the end of the unit read them through slowly.

- Energy may be **transferred** from one place to another, or **transformed** from one type to another, but it is not created or destroyed.
- We use a voltmeter to measure voltage or **potential difference**.
- A high potential difference across a component means that the component is transferring a lot of energy from each bit of charge passing through. A bulb glows more brightly with a high potential difference across it.
- We can use models to help us understand voltage, current and energy in electrical circuits.
- Electricity is generated at power stations. Many power stations burn fuels, which heat water to form steam. Chemical energy in the fuel is transformed to kinetic energy in the steam. The moving steam turns a **turbine**, and this turns a **generator**. In the generator, kinetic energy is transformed to electrical energy.
- Alternative energy sources can be used to turn the generator.
- We use devices to transform and transfer electrical energy to types of energy that are useful to us. Most devices also give out other types of energy, which reduces their **efficiency**.
- Energy is eventually spread throughout the surroundings as heat. We say the energy has been **dissipated**. We cannot use this energy.

Energy changes

○ **How is energy involved in doing useful things?**

Energy makes things happen. In science we say energy is needed to do work. When work is done, two things can happen to the energy:

○ it is **transferred** from one place to another – it moves
○ it is **transformed** from one type of energy to another – it changes form.

You have met energy transfers before. The table opposite shows some types of energy.

Some of these types of energy can be stored.

○ Potential (gravitational) energy is energy stored in something high up. The energy will be changed to kinetic energy if the object falls.
○ Potential (elastic) energy is stored in something that is stretched or twisted, such as an elastic band. If you let go, the stored energy changes to kinetic (and sound) energy as the elastic band snaps back.
○ Chemical energy is stored in batteries and also in food.

Here are some examples of energy transfers and transformations.

Type of energy	Example
thermal (heat)	
light	
sound	
kinetic (movement)	
potential (gravitational)	
potential (elastic)	
electrical	
chemical	

1 Electric light bulb

The transfer: Energy in the wire transfers to the bulb and then to the surrounding air.

The transformations: Electrical energy in the wire transforms into heat and light energy in the bulb.

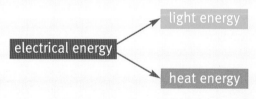

2 Clockwork alarm clock

The transfer: Energy in the coiled spring transfers to the hands and the hammer and bell.

The transformations: Potential (elastic) energy in the spring changes to kinetic energy in the hands and hammer, and then to sound energy in the bell.

potential (elastic) energy → kinetic energy → sound energy

3 Eating a meal

The transfer: Energy in the food is transferred to the girl. This is then transferred around her body and eventually to the air around her.

The transformations: Chemical energy in the food is released in respiration and transformed into kinetic energy (so she can move), heat energy (to keep her warm) and many other forms of energy such as sound (so she can talk or sing!).

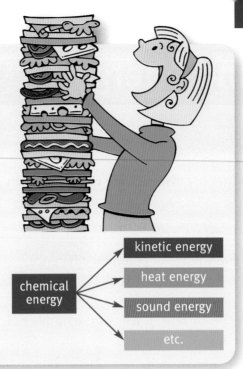

chemical energy → kinetic energy
chemical energy → heat energy
chemical energy → sound energy
chemical energy → etc.

Electrical energy: a useful type of energy

Electrical energy is very useful because it can be transformed into many other types of energy in electrical components. It is transformed into:

o sound energy by a buzzer
o light and heat energy by a bulb
o kinetic energy by a motor.

But there's one snag. We cannot easily store electrical energy. You might think that a battery stores electrical energy, but it actually stores chemical energy. This is transformed into electrical energy when the battery is in a circuit.

In this wind-up radio, you turn the handle to charge it up. Your kinetic energy is transformed to electrical energy in the generator, which is stored as chemical energy in the battery. When you turn on the radio this chemical energy is transformed to electrical energy and then sound energy so you hear the radio.

1 Copy and complete using words from the Language bank:

Energy is needed to do _____ . There are lots of forms of energy including heat, light, sound, potential, kinetic, electrical and _____. Electrical devices transform _____ energy into other forms of energy.

2 Give three examples of an object that has stored energy.

3 Identify the energy transfers and transformations in the following:
   a a whistling kettle on a gas stove
   b a TV
   c a digital camera.

Language bank

chemical
components
electrical
kinetic
light
potential
sound
thermal (heat)
transferred
transformed
work

Electricity and energy

Look at this circuit. You know that we use symbols in a circuit diagram to represent an electrical circuit.

Volts and amps

○ We use a voltmeter to measure the **voltage** or **potential difference** across part of a circuit. It tells us the number of **volts** (V). We connect it in parallel.

○ We use an ammeter to measure the **current**. It tells us the number of **amperes** (A). We connect it in series.

The voltage or potential difference tells us how much push the battery is giving. This push pumps charge around the circuit. The charge moving around the circuit makes the current.

Batteries contain stored chemical energy, which is transformed to electrical energy (the 'push') when the circuit is connected up. The electrical energy is carried by the moving charge. When the charge goes through a bulb its energy is transformed to heat and light energy in the filament.

Modelling electricity and energy

An ammeter measures current, which is how much charge is passing a certain point per second. A voltmeter tells us more about the energy in the circuit.

If you add more batteries in series, the voltage of the circuit becomes higher. The higher the voltage, the more energy is carried by each bit of charge, so bulbs shine brighter and motors spin faster.

The energy is transferred and transformed, but the current is not used up in the circuit.

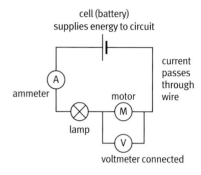

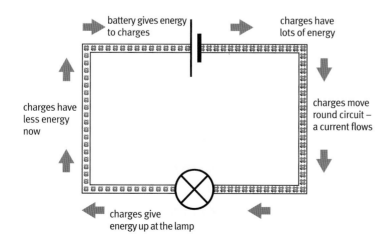

Remember

The circuit must be complete for it to work.

A series circuit is one loop with no branches.

A parallel circuit has branches so the current can split up.

In a series circuit the current is the same through all the components, but the voltage may be different.

Batteries in series add together to give a higher voltage, but in parallel they do not.

The charge travels round the circuit, giving up its energy to the components, but the charge itself is not used up.

The ups and downs of skiing

Try thinking of an electrical circuit as being like this ski slope.

1 The people are like the charge.

2 The number of people going by in a second is like the current.

3 The chair lift lifts the skiers up. It gives them potential energy. It is like the battery, giving the charge energy. The height of the chair lift represents the voltage of the battery.

4 As the skiers go down the slope their potential energy is transformed to kinetic energy. This is like the charge going round the circuit. Its electrical energy is transformed to other types of energy, such as light energy in a bulb or movement energy in a motor.

5 By the time the skiers reach the bottom their potential energy has all been transformed, but the skiers have not been used up. You can still see them! This is like the charge as it gets back to the battery. Its electrical energy needs topping up again but the charge itself has not been used up.

Guess what?

You can make a fruit cell! If you put two different metals such as copper and zinc in a lemon, a current will flow. The two metals provide the voltage, which pushes charge around the circuit and turns the fan.

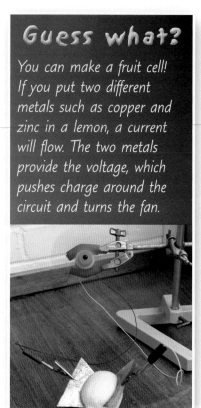

1 Copy and complete using words from the Language bank:

Voltmeters are connected in _____ and ammeters are connected in _____ . The _____ tells us how much 'push' or energy there is in the circuit. The _____ tells us how much charge goes by per second. Electrical energy is _____ in the components of the circuit, but the current is not used up.

2 Explain what voltage and current mean:
   a in the ski slope model **b** in a circuit.

3 In another model for an electrical circuit, you are given some money and you have to go around the shops spending it all before you return home. Relate this to a circuit, including charge, current, voltage and energy. Explain the role of you, the money, the shops and the generous folk who gave you the money.

Language bank

ammeter
ampere (A)
charge
current
energy
parallel
potential difference
series
transformed
volt (V)
voltage
voltmeter

Electricity at home

○ **What are we paying for when we use electricity?**

The national grid

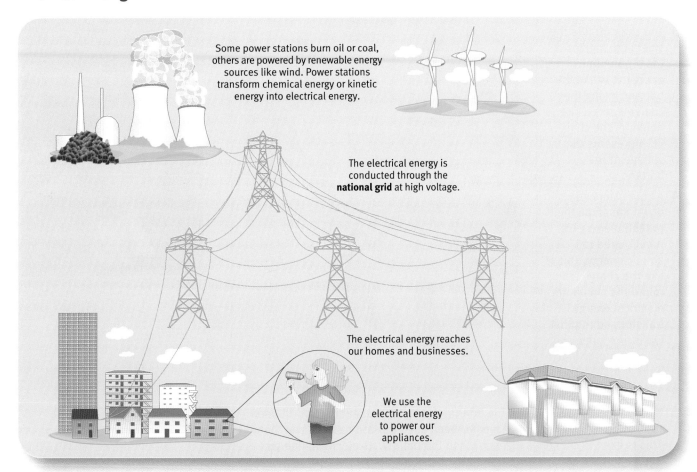

Some power stations burn oil or coal, others are powered by renewable energy sources like wind. Power stations transform chemical energy or kinetic energy into electrical energy.

The electrical energy is conducted through the **national grid** at high voltage.

The electrical energy reaches our homes and businesses.

We use the electrical energy to power our appliances.

Electrical energy is generated at power stations. It then goes into a network of wires called the **national grid**. This transfers it to our homes. Here the electrical energy is transformed to useful energy in appliances such as hair driers and TVs.

The energy you use to run your computer or dry your hair is transferred to your home by the national grid.

Power ratings

When you run an appliance it is using energy. Some appliances transform more energy than others.

The **power rating** of an appliance tells us how quickly it transforms energy. The **power**, measured in **watts** (W), is how many joules of energy per second it uses.

power (in W) = $\dfrac{\text{energy (in J)}}{\text{time (in s)}}$

Look at the photos. A kettle uses much more energy per second than a cassette player.

A 2000 W kettle.

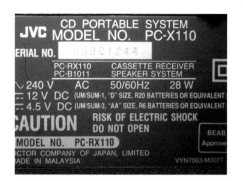

A 28 W cassette recorder.

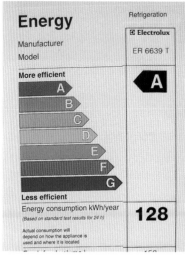

Shops make it easier by rating appliances A to G. The A-rated appliances are the most efficient – they use the least electrical energy to do the job. What rating is your washing machine or fridge at home?

Paying for electrical energy

Electricity companies charge for electrical energy in units called **kilowatt hours**. If you leave a 1000 W (1 kW) heater on for one hour, it will use 1 kW h or one unit of electricity.

National Electric

Mrs I Spy
20 Upper Downlands
Hillmount
OZ2 6XP

Bill period from 19 Mar 2004 to 18 June 2004
Your customer account number
0350293 235898
Electricity

Reading last time	Reading this time	Tariff C – Customer reading E - Estimated reading	Units	Price of each unit in pence	Amount £ p
Meter number(s)		Z76C057255			
73075	75191		2116	6.50	137.54
		Service charge at 10.870p for 92 day(s)			10.00
		Total this invoice			147.54

In this simplified bill, one unit costs 6.5 pence. Mrs Spy's family have used 2116 units since their last bill.

2116 × £0.065 = £137.54.

They must pay a service charge of £10 as well.

1 Copy and complete using words from the Language bank:

_____ is generated in _____ . The energy goes into the _____ which delivers it to our homes. Here it is _____ to useful energy.

2 Look at the power rating on a light bulb and a heater. Which uses more electrical energy per second?

3 The Top Class Electricity Company charges 6.2p per unit and the standing charge is £2. How much would these bills be?
   a Jules uses 100 units   b Mark uses 47.2 units

4 Mervyn uses a 15 W CD player for half an hour. How much electrical energy does it transform in this time?

Can we use an ammeter to measure the power of an appliance?

No, we need a joulemeter to measure the energy it uses. Current × voltage tells us the power, if you really want to know.

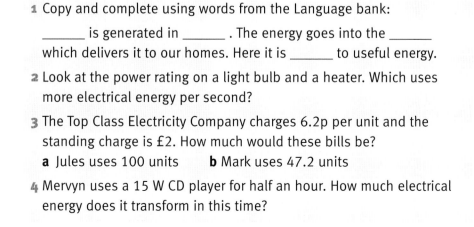

Language bank

electrical appliance
electrical energy
generated
kilowatt (kW)
kilowatt hour (kW h)
national grid
power
power rating
power stations
transformed
watt (W)

The generation game

Where do we get electricity from?

> We can't store the electrical energy from a power station. We have to transform it into other types of energy.

Generating electricity

You know that power stations transform the chemical energy in coal or oil, or the kinetic energy of the wind, into electrical energy. But how does this transformation happen?

It happens inside a **generator**. A magnet spins between coils of wire. This makes an electric current flow in the coils, generating electricity.

Spinning the magnet . . .

All power stations have generators. But they use different ways of turning the magnet.

. . . by burning petrol

The mobile generator on the right is driven by a small petrol engine. The shaft of the engine turns the generator to produce the electricity. In other words, chemical energy in the petrol is changed into kinetic energy and then into electrical energy.

The faster the petrol generator turns, the more kilowatts of power it produces. More power means you can connect up more devices.

. . . in a fossil fuel power station

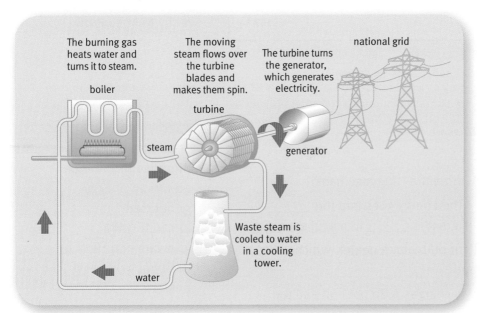

In this fossil fuel power station the fuel has stored chemical energy, which is transformed to heat energy when it burns. The heat energy is used to heat up water to make steam. The steam drives a **turbine** (a big set of blades). The turbine turns the generator.

This power station is at Deeside, North Wales. It burns gas to generate electricty.

. . . in a wind turbine

Wind farms are power stations as well. They convert the kinetic energy that the wind has into electrical energy. The wind turns the blades, and these turn the generator.

The wind is caused by pressure changes in the atmosphere, and by local heating and cooling which cause convection currents to form.

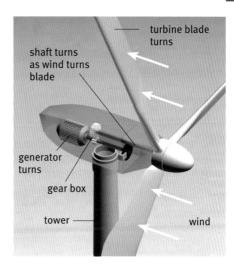

shaft turns as wind turns blade

turbine blade turns

generator turns

gear box

tower

wind

. . . in a nuclear power station

Nuclear reactors do not burn fuel like fossil fuel power stations. Special reactions in some atoms produce massive amounts of heat. The heat is used to make steam, which drives the turbines.

Talking about a generation problem

Burning fossil fuels to generate electricity damages the environment. Carbon dioxide is produced, which adds to the greenhouse effect. This causing global warming, which could be catastrophic if some predictions come true.

Impurities in fossil fuels burn to produce sulphur dioxide and oxides of nitrogen, and these cause acid rain. At some power stations these acidic gases are removed before the waste gases go into the air.

What can you do?

How many times have you heard . . .

If you are not using it, switch it off!

Guess what?

Electricity is produced on demand (as it is needed), as we cannot easily store it. Massive peaks in demand sometimes happen during commercial breaks on TV as people rush to the kettle to make a drink.

Guess what?

If everyone just boiled the amount of water they needed, we could close down at least one power station.

1 Copy and complete using words from the Language bank:

Electricity is made in a device called a _____ . This transforms kinetic (movement) energy into _____ . A fossil fuel is burned in some power stations. This heats water to form _____ , which drives the _____ which turns the generator.

2 Explain how a wind turbine generates electricity.

3 Fossil fuels were formed from dead animals and plants millions of years ago. Explain the following statements.
   a Electrical energy from fossil fuel power stations originally came from the Sun.
   b Electrical energy from wind turbines originally came from the Sun.

4 Some wind turbines are out at sea. Find out some advantages and disadvantages of having them at sea rather than on land.

Language bank

acid rain
coil
electrical energy
fossil fuel power station
generator
global warming
greenhouse effect
nuclear power station
steam
turbine
wind turbine

113

Wasting energy

○ How can we reduce the waste of energy?

A light bulb transforms electrical energy to light energy and heat energy. You can feel this heat energy if you put your hand near the bulb (carefully!). This heat energy is not useful, because what we wanted from the light bulb was light energy.

In an ideal world, a light bulb would transform all the electrical energy to light energy. This would be an efficient process and no energy would be wasted. But in reality, energy transformations are not 100% efficient. Some energy is always wasted.

What happens to the wasted energy?

In an energy transformation, there is the same amount of energy at the end as there was at the beginning. Energy is **conserved**.

If energy is conserved, how can it be wasted? The problem is that it is often transformed to heat energy, and this is not useful. It spreads through the surroundings, warming them up just a tiny bit. When energy is spread out we say it is **dissipated**. We cannot collect up and use the energy for anything once it has been dissipated.

Look at the diagram of generating electricity in a fossil fuel power station on page 112 . There are several energy transformations:

chemical energy → heat energy → kinetic energy → electrical energy

Each transformation wastes energy, giving an overall efficiency of 30–40%. Just think how much energy is dissipated in cooling towers.

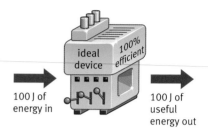

100 J of energy in

100 J of useful energy out

In an ideal device, no energy is wasted.

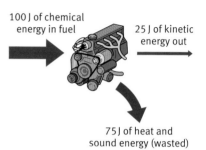

100 J of chemical energy in fuel

25 J of kinetic energy out

75 J of heat and sound energy (wasted)

Real machines like this engine waste energy. Only 25% of the chemical energy in the fuel turns into useful kinetic energy. The rest is dissipated as heat and sound energy.

Energy-efficient light bulbs

Energy-efficient bulbs are made of a fluorescent tube containing a gas that glows when electricity passes through it. Traditional bulbs have a hot tungsten wire which glows white-hot to give out the light. This is less efficient.

An ordinary 100 W light bulb is only 20% efficient. The other 80% of the energy is dissipated as heat energy. The diagram compares this with an energy-efficient light bulb.

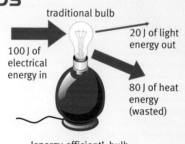

Both bulbs give the same amount of light energy. The energy-efficient bulb wastes a lot less energy.

traditional bulb

100 J of electrical energy in

20 J of light energy out

80 J of heat energy (wasted)

'energy-efficient' bulb

26 J of electrical energy in

20 J of light energy out

6 J of heat energy

Bright idea quiz

Why should we use energy-efficient light bulbs? (Answer A–E)

A They use less electricity than traditional bulbs.

B They save you money. Less electricity costs less.

C They give just as much light.

D They last longer and don't have to be replaced as often.

E All of the above.

The correct answer is E of course!!

ARE ELECTRIC VEHICLES EFFICIENT?

Instead of burning fuel, **electric vehicles (EVs)** use an electric motor powered by batteries to move the vehicle along. They can travel 100 miles or so before they need charging up again, and the charging takes 5–6 hours.

This EV has an electric motor.

Fuel cell cars are another recent development. In a fuel cell, a chemical reaction converts hydrogen and oxygen to water. The reaction also gives out lots of electrical energy, and this powers a motor which drives the car along.

Electric vehicles and fuel cell cars give out no harmful exhaust emissions at all, so they are a 'green' alternative to petrol- and diesel-powered cars.

But are electric vehicles and fuel cell cars the most efficient way of travelling? Let's look at the figures for some typical vehicles.

This car is powered by a fuel cell.

Type of car	Process	Energy transformation	Efficiency of process	Overall efficiency of car
petrol	burning petrol in internal combustion engine	chemical energy → kinetic energy	20%	20%
electric	1 battery powers the motor	chemical energy → electrical energy	90%	72%
	2 motor moves the car	electrical energy → kinetic energy	80%	
fuel cell	1 making hydrogen from methanol	chemical energy → chemical energy	30–40%	24–32%
	2 fuel cell reaction drives motor	chemical energy → kinetic energy	80%	

Energy is wasted in all the transformations. But the electric car seems to have the most efficient method of producing kinetic energy.

However, the batteries have to be charged up with electrical energy. This may come from a fossil fuel power station, which is no more than 40% efficient. This makes the electric car only about 28% efficient – not very much better than a petrol car engine. But hydroelectric power stations are more efficient than fossil fuel power stations. Powered this way, electric vehicles become 65% efficient.

Electric cars concentrate the pollution at power stations, so help clean up city centres.

1 Copy and complete using words from the Language bank:

Machines carry out _____, which are never 100% efficient. Some energy is always wasted. The wasted energy is spread through the surroundings and _____.

2 **a** If energy is conserved in transformations, how can it be wasted?

b Explain how an energy-efficient light bulb can give out the same amount of light as a traditional bulb but use less electrical energy.

3 Why do electric cars become more efficient when charged up on hydroelectric power rather than electricity from a coal-fired power station?

Language bank

conserved
dissipated
generating electricity
efficiency
efficient
electric vehicles
energy transformations
fuel cell cars
wasted energy

115

Checkpoint

1 Choose the answer

Decide whether the following statements are true or false. Write out the true ones. Correct the false ones before you write them down.

a Electrical energy is useful because it can be transferred to different places along wires.

b Electrical energy is useful because you can store it.

c Electrical energy is useful because it can be transformed to many different types of energy by devices and components.

d Energy is always lost when it is transferred or transformed.

2 Down the slope

Copy each sentence below about the ski slope model of an electrical circuit. Complete each one using one of the following words:
current, potential difference, charge, energy, component, battery

a The chair lift is like the . . .

b The chair lift gives the people . . .

c The height of the chair lift represents the . . .

d The people are like the . . .

e The number of people going by per second represents the . . .

f At the bottom of the hill, the people have run out of . . .

3 Power stations

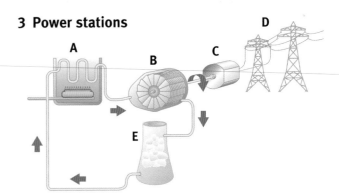

The diagram above shows a power station.

a Choose the correct label from the following list for A to E:
generator, boiler, cooling tower, turbine, national grid

b Choose the correct energy transfer or transformation below for each label A to D:
kinetic energy transferred
chemical energy → heat energy → kinetic energy
electrical energy transferred
kinetic energy → electrical energy

c Where does most energy become dissipated, at A, B, C, D, or E?

4 Alternative energy resources

Copy and complete the following sentences, choosing the correct words.

a Fossil fuel power stations produce carbon dioxide which may cause **acid rain / global warming / the ozone hole**.

b Fossil fuel power stations produce acidic gases such as sulphur dioxide which may cause **acid rain /global warming / the ozone hole**.

c Alternative energy resources can be used to **generate /store / save** electricity.

d Alternative energy resources use different methods of **burning the fuel / turning the generator / transferring the electrical energy**.

Gravity and space

Before starting this unit, you should already be familiar with these ideas from earlier work.

○ In our Solar System, the Earth is one of nine planets orbiting our star the Sun. What is the difference between the way we see a star and a planet?

○ As the Moon orbits the Earth, its appearance changes as a different view of it is lit up. What do we call this?

○ An eclipse happens when a shadow hides our view of something.

○ A force changes the shape, direction or speed of something. A force has magnitude and direction.

○ Mass is the amount of matter in an object. Weight is a force, which is caused by gravity acting on the object. What units do we use to measure mass and weight?

You will meet these key ideas as you work through this unit. Have a quick look now, and at the end of the unit read them through slowly.

○ Gravity is a force of attraction that acts between objects that have mass. The magnitude of the force depends on the mass of the objects and the distance between them.

○ The Earth has a big mass so it exerts a large gravitational force. But the force gets less as you travel away from the Earth's surface into space.

○ Different planets have very different sizes and masses. This means they exert different gravitational forces.

○ Weight is the force of gravity acting on an object's mass. So the weight of an object would change on different planets, depending on the gravitational force at the surface of each planet.

○ The planets in the Solar System have different masses and are in orbits at different distances from the Sun. So they each experience a different attraction to the Sun, which keeps them in their individual **orbits**.

○ We put artificial **satellites** into orbit around the Earth. They have different orbits depending on their use, and are kept in orbit by the pull of Earth's gravity.

Gravity

○ What is gravity?

This roller coaster climbs to a height of around 130 m. Then gravity pulls it back down to Earth, reaching a terrifying 120 miles per hour in just 4 seconds.

Gravity is a force which affects you every day. It keeps you sitting on your chair and makes things feel heavy. Like a magnet, gravity can act between objects that are not even touching – it is a non-contact force.

A magnetic force attracts a magnet to a magnetic material. Gravity is different. It pulls together any objects that have **mass**. The more mass they have, the bigger will be the **gravitational force** between them.

The Earth has a huge mass. It exerts a huge gravitational force. So any object that has mass, on the surface of the Earth or close to it, is attracted to it. Wherever you are on Earth, gravity acts to pull you towards the centre of the Earth.

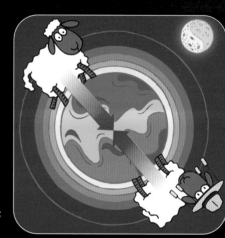

Gravity attracts the sheep strongly towards the centre of the Earth.

A two-way street

But gravity is a two-way thing. The sheep have mass too, and they exert their own gravitational force, pulling the Earth towards them. Because the Earth's mass is so much bigger, you don't notice the tiny effect that the sheep have on the Earth.

The Moon moves in a shape called an **ellipse** (an oval), in **orbit** around the Earth. The gravitational force of the Earth attracts the Moon and keeps it moving in its path around the Earth.

But the Moon also attracts the Earth. This has several effects that we notice on Earth, particularly the height of the oceans and seas. The Moon's gravity causes the tides.

A little bit of history

It is often said that Newton discovered gravity. The story goes that while he was sitting thinking under an apple tree, an apple fell off the tree and suddenly gravity was discovered.

The reality is that Newton put together some mathematics. His calculations used gravity as a central idea to explain why apples fall and planets move in orbits. The falling apple story may or may not be true, but it is not the whole story.

Mass and weight

Every object has mass. So it is attracted to the Earth's centre by gravity. The gravity causes a mass to have **weight**.

Gravity on Earth pulls each gram of mass with a force of about 0.01 newton. Put another way, a 1 g mass has a weight of 0.01 N.

The more grams there are, the bigger the weight.

- A mass of 100 g (an average apple) is attracted by a force of 1 N: its weight is 1 N.
- A mass of 1000 g (1 kg, or a bag of flour) is attracted by a force of 10 N: its weight is 10 N.

We say that the **gravitational field strength** on Earth is 10 newtons per kilogram, or 10 N/kg. (We could write this as 0.01 N/g instead.)

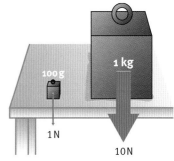

The larger the mass of the object, the bigger the force that attracts it towards the centre of the Earth.

Working out weight

To work out the weight of an object, just multiply the mass in kilograms by 10. This gives you the weight in newtons.

Object	Mass	Weight (mass in kg x 10)
bag of flour	1 kg	10 N
bag of sugar	0.5 kg	5 N
interesting science book	0.2 kg	2 N
sheep	85 kg	850 N

Don't confuse mass with weight. Mass tells us how much matter is in the object, and it doesn't change. Weight depends on the force of gravity.

1 Copy and complete using words from the Language bank:

   Gravity is a _____ which acts between objects that have _____.
   Gravity acts on Earth towards the _____ of the planet.

2 Why is the force of gravity between the Earth and a sheep larger than the force between the Earth and a bag of sugar?

3 Calculate the weight of the following objects:
   a a 2.5 kg bag of potatoes
   b a 75 kg teacher.

4 Peter thinks that gravity pulls things downwards. While he is not completely wrong, explain what he has failed to realise.

Language bank

centre
force
gravitational field strength
gravitational force
gravity
kilogram
mass
newton
weight

Changing gravity

○ How does gravity change?

Where in the world are you?

You know that the force of gravity depends on the mass of an object – more mass means more force. The force of gravity also depends on where things are. If two large objects are close together there is a strong force of gravity pulling them towards each other. But if they are very far apart, then the force of gravity between them is smaller.

These astronauts in the International Space Station are in orbit 400 km above the Earth. The Earth's gravity holds their spacecraft in orbit. It also holds them in orbit. But because they are travelling along the same path as their spacecraft, they can float about inside it.

The further you go from Earth, the weaker its gravitational force becomes. Go millions of miles from Earth and its attractive force is so weak it's hardly noticeable. Spacecraft a long way from Earth or any other planets, that feel no gravity, are 'weightless'.

Guess what?

The Sun is larger than the Earth – it has a much bigger mass. So even though it is a very long way away, its force of gravity still has an effect. The Sun's gravity attracts all the planets of the Solar System and keeps them in orbit.

Gravitational attraction keeps Earth orbiting the Sun.

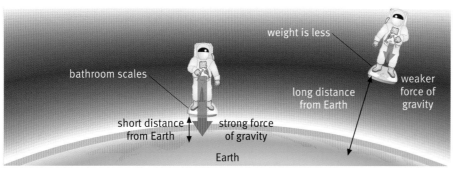

The further away from Earth you are, the less you weigh.

Walking on the Moon

The Moon is much smaller than the Earth. Gravity on the Moon is about one-sixth of that on the Earth. So astronauts walking on the Moon weigh less than they do on Earth.

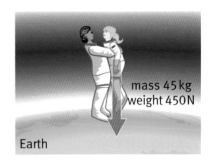

mass 45 kg
weight 450 N

Earth

A 1 kg mass on the Earth has a weight of 10 N. But on the Moon it would weigh about one-sixth as much, 1.7 N.

Let's say your mass is 45 kg. On Earth your weight is 450 N. This means someone would need a force of at least 450 N to lift you up. But on the Moon you weigh only one-sixth of this, 75 N – much easier!

Why do you weigh less on the Moon?

The Moon has a smaller mass than the Earth. So its force of gravity is less. We say the Moon has a smaller gravitational field strength.

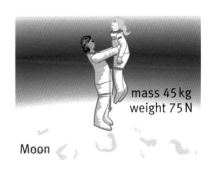

mass 45 kg
weight 75 N

Moon

Your mass does not change if you go to the Moon. It's the force pulling you towards the centre of the Moon that is smaller. So your weight there is less.

What is your weight on other planets?

All the planets in our Solar System have different sizes. So their gravitational field strengths are different. Pluto has a small mass while Jupiter and Neptune are huge. If you could visit these planets your mass would not change, but your weight would.

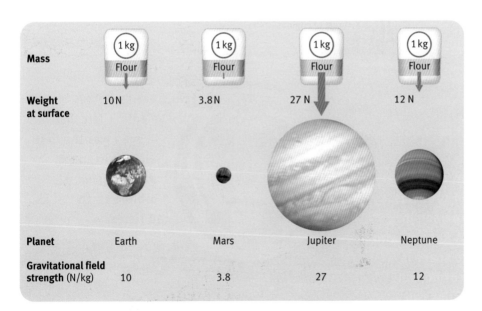

Mass	1 kg Flour	1 kg Flour	1 kg Flour	1 kg Flour
Weight at surface	10 N	3.8 N	27 N	12 N
Planet	Earth	Mars	Jupiter	Neptune
Gravitational field strength (N/kg)	10	3.8	27	12

1 Copy and complete using words from the Language bank:

The force of gravity depends on mass and _____. The larger the mass of the object, the stronger the force of _____. The further away objects are from each other, the weaker the _____ of attraction.

2 An 85 kg sheep weighs 850 N on Earth. What would it weigh:

a on the Moon   **b** on Neptune?

3 Describe why you might have trouble getting out of bed on Jupiter.

4 Imagine how life on the Moon might be different from living here on Earth. Describe some differences you would notice.

Language bank

attraction
distance
force
gravitational field
  strength
gravity
mass
weight

More about gravity

Taking off

The American space agency NASA fires rockets into space on a regular basis. They go up into orbit. In order to leave the Earth's surface a rocket needs to overcome its weight.

Weight is the force of gravity pulling the mass back down towards the Earth. The powerful engines produce enough thrust to lift the rocket off the ground and into space. The thrust therefore acts against gravity.

1 The thrust of the engines needs to be larger than the rocket's weight for take-off.

2 As the rocket gets higher, force of gravity (its weight) gets less and so less thrust is needed to overcome it.

3 The rocket sheds the large fuel tank and solid booster rockets, so it loses mass. This again means that less thrust is needed.

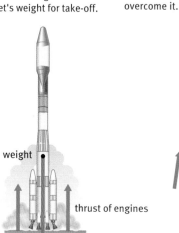

weight

thrust of engines

Interesting facts about space exploration

The Russian Yuri Gagarin in 1961 was the first person in space. (A Russian who goes into space is called a cosmonaut; other countries call them astronauts.)

Valentina Tereshkova in 1963 was the first woman in space. Also Russian, she orbited Earth 48 times.

Helen Sharman in 1991 was the first British woman in space.

The Americans Neil Armstrong and Buzz Aldrin in 1969 were the first to walk on the Moon.

Guess what?
Only 12 people have ever walked on the Moon.

A successful failure: using gravity

In 1976, an accident happened while the Apollo 13 spacecraft was on its way to the Moon. NASA used the gravitational attraction of the Moon to get the damaged Apollo 13 spacecraft back to Earth.

Apollo spacecraft had three sections:

- The command module carried the three astronauts in space.
- The service module carried fuel and had engines to provide thrust, to adjust course once away from the Earth's gravity.
- The lunar module was to take two astronauts to land on the Moon, while the third stayed in orbit around the Moon in the command module.

But an explosion destroyed part of the service module. The mission became a fight for survival for the astronauts on board. They used the lunar module as a 'lifeboat'. The Moon's gravity attracted the spacecraft, which orbited the Moon.

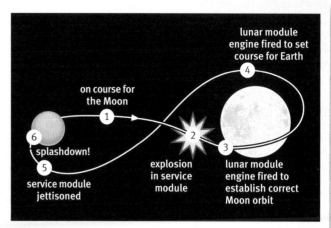

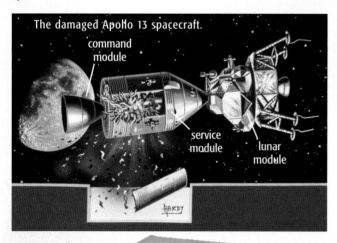

The damaged Apollo 13 spacecraft.
command module
service module
lunar module

Once it was travelling in the right direction they used a short thrust from the lunar module's engine to overcome the Moon's gravity and get back to Earth.

The astronauts had to move back to the command module for splashdown as that was the only part that could land safely.

Splashdown.

1 Copy and complete using words from the Language bank:

Rockets use a force called _____ to overcome their _____. As a rocket moves away from the Earth the force of _____ gets smaller. Rockets often become lighter as they use fuel and shed body parts. This also means that the force of _____ decreases.

2 Explain the function of the three parts of the Apollo 13 spacecraft.

3 Find out about NASA's more recent space missions by typing 'NASA' into a search engine.

Language bank

gravity
mass
orbit
thrust
weight

The story of the Solar System

○ **How have our ideas about the Solar System changed?**

Imagine you are a cave dweller looking at the night sky thousands of years ago. You wonder why the stars and planets seem to move across the sky. You don't know what causes day and night, or the seasons.

There have been many theories to explain all this. Some interesting ideas about the Solar System have developed.

Ideas about the Solar System

Ancient times

Many ancient civilisations had cultures that were based on myths, legends and superstitions. The reasons that people gave for things happening had no scientific evidence or reasoning to support them.

Eventually people tried to explain events like day and night, and observed that stars and planets moved in a regular way across the sky. The Babylonians even predicted an eclipse, but still considered celestial bodies (planets and stars) to be gods.

The Egyptians, the Chinese and others had star maps which they used to place festivals on the calendar and to work out when to sow and when to harvest their crops. The Ancient Greeks used stars to navigate.

Eclipses were terrifying events for people who had no idea why the Sun suddenly stopped shining.

> **Remember**
> The **Solar System** is our Sun (a star) along with the planets and other bodies that orbit it.

Science emerges

Thales was one of the first Greek scientists who tried to find physical explanations for things they saw. Thales thought the Earth floated in water while others believed it was a disc suspended freely in space. Pythagoras and his followers thought the Earth was a sphere which rotated round a central fire (but they did not identify this as the Sun).

Aristotle suggested that the Earth was at the centre of the universe. This was later supported by Ptolemy. He produced the **geocentric model** of the Solar System shown here.

Aristarchus of Samos did not agree. He suggested that the Earth turned on its own axis every 24 hours, and that it moved around the Sun along with other planets. Most people did not take his ideas seriously.

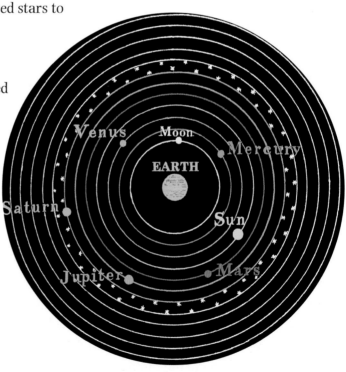

The geocentric model suggested that the Earth was fixed at the centre of the universe. In this model the planets that were known at that time and the Moon all orbited the Earth in perfect circles.

A scientific revolution

The Earth-centred model lasted for about a thousand years. But Leonardo da Vinci and others questioned the idea of an immobile Earth at the centre of the universe.

The sixteenth-century scientist Copernicus explained the movement of planets by having the Sun in the centre rather than the Earth. This was known as the **heliocentric model** of the Solar System. Again people did not take his views very seriously.

Galileo had hard evidence to suggest that Copernicus' ideas were true. With his telescope he observed that planets like Venus did not orbit the Earth. He was warned by the authorities not to promote his ideas.

Kepler used many people's observations and measurements to devise laws of motion for the planets. He showed that they moved not in circles but in elliptical (oval) orbits.

Newton argued that it was the attractive gravitational force between the Sun and each of the planets which keeps them in motion.

Our modern-day view of the motion of the planets is based on the work of Newton. The planets are held in elliptical orbits by the gravitational attraction of the Sun.

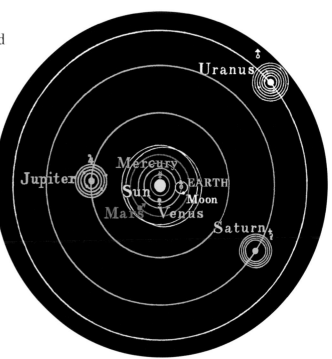

The heliocentric model had the Sun at the centre of the universe with the Earth and other planets moving round it.

Newton also realised that a force must exist between an orbiting object such as the Moon, and the Earth. But the Moon is also moving forwards. So it ends up in an orbit around the Earth, a combination of falling and moving forwards.

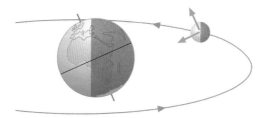

Language bank

Aristarchus of Samos
Aristotle
Copernicus
Earth
elliptical
Galileo
geocentric model
heliocentric model
Newton
orbit
planets
Ptolemy
Solar System
Sun
Thales

1 Copy and complete using words from the Language bank:

People first believed that the Earth was the centre of the _____, and that everything rotated around it. This was known as the _____. Much later, others suggested that the _____ was the centre of the Solar System. This was known as the _____ .

2 What was Kepler's contribution to the theory of the Solar System?

3 Explain the modern-day view of the Solar System.

4 Find out more about why people did not believe Galileo, and the problems he faced.

○ **What keeps the planets and satellites in orbit?**

We think the Sun has a mass of about 2 million million million million million kilograms (that's 2 with 30 zeroes after it!). So it exerts a huge gravitational force. This keeps all the planets in the Solar System, including the Earth, in orbit around it.

The Earth and some other planets also have moons and other bodies in orbit around them, kept there by the planets' gravity.

> What keeps the planets and moons orbiting in different positions?

> Each experiences a different force of gravity, depending on its mass and its distance from the Sun.

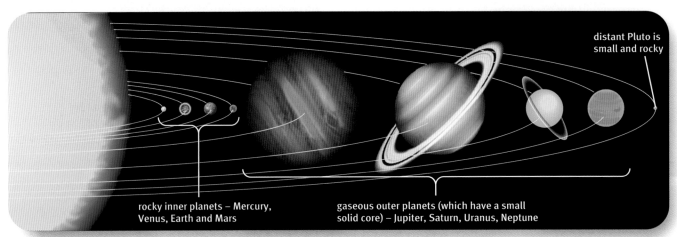

distant Pluto is small and rocky

rocky inner planets – Mercury, Venus, Earth and Mars

gaseous outer planets (which have a small solid core) – Jupiter, Saturn, Uranus, Neptune

The Solar System, showing the sizes of the planets and the Sun. The distances are not to scale.

Circular motion

Imagine swinging a bucket of water on a rope around your head. The water stays in the bucket – as long as you keep holding the rope. If the rope breaks, the bucket flies off in a straight line.

The tension in the rope is like the force of gravity holding the planets in orbit round the Sun, or the Moon in orbit round the Earth.

Our satellite Moon

An object that orbits another body is called a **satellite**. The Moon is a natural satellite, orbiting the Earth.

pull of rope

pull of bucket

With no pull from the rope, the bucket flies off in a straight line.

When the rope breaks, there's no force keeping the bucket in orbit and it goes off in a straight line. The same would happen to the planets without the Sun's gravity.

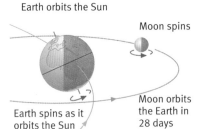

Earth orbits the Sun

Moon spins

Earth spins as it orbits the Sun

Moon orbits the Earth in 28 days

The Moon takes 28 days to spin round once on its axis. It also takes 28 days to orbit the Earth. So we never see the far side of the Moon.

Why does the Moon orbit the Earth?

There are two current theories.

The collision hypothesis (or the giant impact theory)

The theory: The Earth collided with another planet (about the size of Mars) and threw off a huge mass of hot gas. This cooled and formed the Moon.

Evidence to support this: The mineral composition of the Moon is similar to that of Earth.

The Moon has a small core which is thought to be similar to Earth's.

The capture hypothesis

The theory: The Moon was a large asteroid (a space rock) that was captured by the gravitational attraction of the Earth.

Evidence to support this: Relative to the size of the Moon, the Moon's core is much smaller than that of the Earth and its crust makes up a larger proportion. The Moon's core contains less metallic material.

Artificial satellites

Artificial satellites are objects put in orbit around the Earth by people. There are different types of orbit as the diagram shows. **Geostationary** satellites orbit at the same speed as the Earth's spin. We use satellites for:

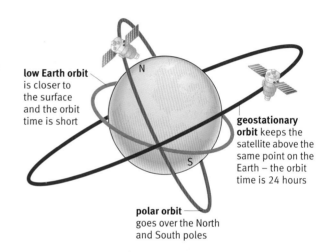

low Earth orbit is closer to the surface and the orbit time is short

geostationary orbit keeps the satellite above the same point on the Earth – the orbit time is 24 hours

polar orbit goes over the North and South poles

o **Communications:** to transmit radio, TV and telephone signals around the world. Geostationary satellites are used for this, each one transmitting signals to an area below it.

o **Navigation:** GPS (global positioning system) helps sailors, walkers and even car drivers to locate their exact position. There are 24 satellites around the Earth.

o **Weather:** satellites such as Metsat are used to observe the Earth's weather patterns. This helps forecasters to predict the weather.

o **Observing the Earth:** military satellites are used for spying. They can photograph anything from a country's weapons to the number of sheep in a field.

o **Exploring the universe:** the Hubble telescope is outside the Earth's atmosphere so it can take better photos of the rest of the universe.

1 Copy and complete using words from the Language bank:

_____ keeps the planets orbiting the Sun and keeps the _____ orbiting the Earth. Without gravity the _____ would travel away from the Earth in a straight line.

2 What is a satellite?

3 a Give an example of how artificial satellites are used.
   b Why are communications satellites put in geostationary orbits?

4 Describe the two main theories about why we have the Moon.

Language bank

capture hypothesis
collision hypothesis
Earth
geostationary orbit
GPS
gravity
Hubble telescope
Moon
navigation
orbit
planets
satellite
Solar System
Sun

Checkpoint

1 All about gravity

Copy and complete the following sentences, choosing the correct words.

a Gravity is a **direction / distance / force**.

b Gravity **attracts things to each other / only comes from Earth / acts mainly on magnetic materials**.

c Gravity gets bigger if the **volume / mass / distance** is bigger.

d Gravity gets smaller if the **volume / mass / distance** is bigger.

e Your **speed / weight / size** is the pull of gravity on your mass.

2 Watching your weight

Ben and Lina have the same mass. For each situation below, say which of them is heavier (has the bigger weight) and why.

a Ben is on the beach and the Lina is in a spacecraft half-way to the Moon.

b Ben is standing on the surface of the Moon and Lina is having lunch on the school field.

c Ben is still on the Moon and Lina is now having tea on Jupiter.

3 Complete the table

Copy and complete the table below, calculating the missing figures.

4 In orbit

Sketch the diagram below. Choose the correct label for each orbit:

polar orbit

low Earth orbit

geostationary orbit

a Which orbit has the shortest orbit time?

b Which orbit keeps the satellite above the same point on the Earth's surface?

c If gravity stopped working the satellites would **move off in a straight line / fall back down to Earth / crash into each other**.

Planet	Gravitational field strength (N/kg) on the planet	Weight of a 1 kg mass on the planet	Weight of a 50 kg mass
Earth	10	10 N	500 N
Mars	3.8	3.8 N	
Jupiter	27		1350 N
Saturn	9		
Pluto		0.6 N	

Speeding up

Before starting this unit, you should already be familiar with these ideas from earlier work.

- Speed tells us how fast something is moving. What two things do you need to measure to work out speed?
- If the forces on a stationary object are balanced, the object does not move.
- If the forces on a stationary object are unbalanced, the object will start to move.
- If an object is already moving, a force will change the way it is moving. Think of the force of the brakes on a moving car.
- Friction slows down moving objects. What do we call the friction when something is moving through the air?

You will meet these key ideas as you work through this unit. Have a quick look now, and at the end of the unit read them through slowly.

- If an object is moving at a steady speed in a straight line, the forces on it are balanced.
- A moving object will move faster if a force is applied in the direction of its movement. We call a change in speed an **acceleration**.
- **Drag** forces act when an object moves through air or water. The friction is caused by the air or water particles hitting the moving object and slowing it down.
- The force of air resistance depends on the object's shape. A **streamlined** shape causes less drag.
- The force of air resistance also depends on the object's speed. The faster it moves, the more the air particles hit it and the greater the friction.

How fast?

○ How fast is it moving?

What do we mean by speed?

Speed is how long it takes to travel a certain distance. In this race all the women are running 100 m. The one who finishes first (in the shortest time) is the fastest. She is the one with the highest speed.

The speed is not the same throughout the race. The runners speed up off the blocks at the start. There is a spurt at the end for the line. So what is the speed for the race?

If we know the total distance for the race and the total time an athlete took to run it, we can work out the **average speed**. For example, Ken ran 100 m in 9.84 seconds:

$$\text{average speed} = \frac{\text{total distance}}{\text{total time}}$$

$$= \frac{100 \text{ m}}{9.84 \text{ s}} = 10.2 \text{ m/s}$$

Just remember SIDOT, Speed Is Distance Over Time.

This table shows the times for five races. You divide the distance by the time to find the average speed. Convert the time to seconds first.

Race distance (m)	Best men's time	Best women's time	Average speed for men (m/s)	Average speed for women (m/s)
100	9.84 s	10.49 s	10.2	9.5
200	19.32 s	21.34 s	10.4	9.4
800	1 min 41.73 s	1 min 53.28 s	7.9	7.1
1500	3 min 27.37 s	3 min 50.46 s	7.2	6.5
10 000	26 min 38.08 s	29 min 31.78 s	6.3	5.6

The average speeds for running five races.

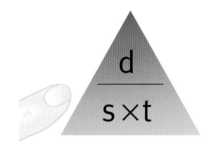

The formula triangle helps you with speed calculations. Put your finger over the letter you want to find out – distance (d), speed (s) or time (t).

Using a stopwatch is less accurate. The person's reaction time might affect the result.

Measuring speed

To calculate an accurate speed we need to measure the time carefully.

The timer used for top class events can time to hundredths of a second. A photo helps decide the results if it's very close.

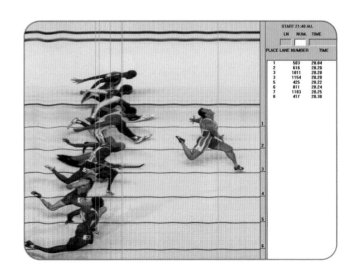

Dividing the total distance by the time tells us the average speed for a journey. But speed varies throughout a journey or race. How do we measure the speed of something at a particular moment? Here are two methods:

This 'speed gun' sends out a radar wave and then detects the wave which bounces back off the cars. The difference between the radar waves sent out and those that come back tells the police officer about the speed.

We measure how wide the card is. The light gate times how long it takes for the card to pass through it. The computer calculates the speed of the trolley as it passes through the light gate. This method is very good for measuring low speeds in the lab.

Changing speed

Acceleration is a change in speed. A car's speed might increase from 0 to 60 miles per hour. It has **accelerated**.

A Formula 1 racing car gets to 60 miles per hour from a standing start much faster than a family car. It has a bigger acceleration.

If you have ever free-wheeled down a hill on a bicycle, you will know that going downhill makes you accelerate.

The racing car accelerates faster than the family car.

1 Copy and complete using words from the Language bank:

How fast something moves is its _____. Average speed is total _____ divided by total _____. It can be measured in different _____ such as metres per _____, miles per _____ or kilometres per _____.

2 Which travels faster, a car that goes 30 miles in 30 minutes or one that goes 20 miles in 35 minutes?

3 If Jo walked at a speed of 8 m/s for 5 minutes, how far did she walk? (Look at the formula triangle.)

4 Explain why we need different methods for measuring speed, such as a radar gun for catching speeding motorists and a stopwatch on sports day.

Language bank

acceleration
average speed
distance
hour
kilometre
light gate
metre
mile
radar
second
speed
stopclock
time
units

How forces can speed you up

○ How do forces affect speed?

You know that a force is a push or a pull. A force can change an object's shape, speed or direction. A change in speed is **acceleration**.

Forces and acceleration
Look at the three photos below. Think about the forces.

1 *The thrust of the engine is speeding the car up. It is accelerating. The forces are unbalanced.*

2 *The skater is moving at a steady speed. The forces are balanced.*

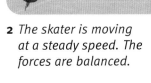

Newton and his laws
Newton used scientific language to describe what we have seen about forces and moving objects.

3 *Nothing is changing shape, speed or direction. The forces are balanced.*

> An object will remain stationary or at uniform (steady) speed, unless subjected to an external force.

A closer look at forces

The bowling pin is stationary because there are no forces.

Yes there are. Gravity (weight) acts downwards. The reaction force of the floor acts upwards. The forces are balanced.

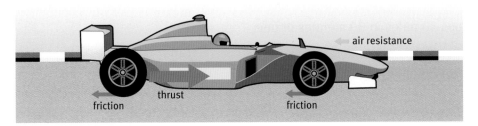

1 *The forward thrust of the engine is bigger than the backwards forces of friction and air resistance, so the car accelerates. The weight of the car is balanced by the reaction force of the ground, but these are not shown on this diagram.*

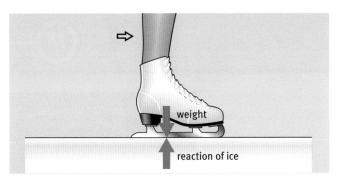

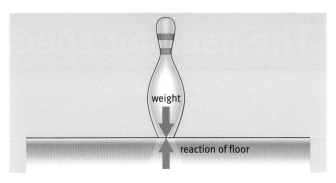

2 *The skater pushed off to get herself moving, but there is no forward force acting at the moment. There is hardly any friction with the ice. She moves at a steady speed. Her weight is balanced by the reaction of the ice.*

3 *The weight of the pin is balanced by the reaction force of the floor so it stays stationary – at least until you bowl!*

Forces get things going

To get something moving you need to change its speed. You need an unbalanced force. The larger the force you use, the faster the object will accelerate. This is why shot-putters are large and strong.

To get a car moving, the engine exerts a force on the car. Bigger engines often give the best acceleration.

The shot-putter exerts a large force to give the shot a big change in speed.

Car	Engine size (litre)	Time taken to accelerate from 0 to 60 m.p.h. (s)
Ford Mondeo	1.8	10.4
Maclaren F1	6.1	3.2
Grandma's old Mini	1.0	22
Porsche 911 Turbo	3.6	5.0

Car magazines often quote '0 to 60 times'. The best acceleration comes from an engine that delivers a big force at the wheels. But mass makes a difference too – making the car lighter improves its acceleration.

1 Copy and complete using words from the Language bank:

A _____ can produce a change in speed, which we call _____. If the forces are balanced, an object will continue moving at a _____ or remain _____ .

2 Carl is cycling along the road at a steady speed.
  a What can you say about his forward cycling force and the forces of friction and air resistance?
  b Name two other forces acting on Carl.

3 An ice hockey puck is hit along the ice. Draw a diagram to show the forces acting on the ice puck as it moves.

4 Shot-putters are usually large and heavy, but sprinters tend to be slim and light. Both need to be muscular. Explain why this is, using the word 'acceleration'.

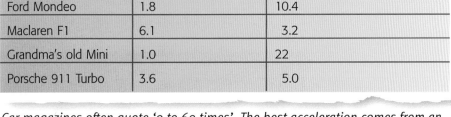

Language bank

acceleration
air resistance
balanced
force
friction
mass
reaction
stationary
steady speed
unbalanced
weight

The need for speed

It's a drag

When you are moving forward, friction with the ground and air resistance slow you down. If you are in water, the water acts to resist you. We call these forces that slow you down **drag**.

Drag happens because as you move forward you knock into the particles in the air or water. They slow you down.

How do we increase speed?

Imagine a new car is being designed and you have been given the job of speeding it up. You cannot change the engine or the car's mass. But you can speed it up if you can reduce the drag.

If the car has a **streamlined** or **aerodynamic** shape, it will cut through the air like a dart. Aerodynamic shapes have less surface area for the air particles to push against as they move. Even lorry designers try to make the shape more aerodynamic by using wind deflectors.

Modern cars have become more wedge-shaped and less square. This makes them more aerodynamic and reduces the drag forces.

This car is in a wind tunnel. The smoke shows how the air flows over the car.

The Model T Ford, 1908.

The Lamborghini Diablo, 2003.

Every second counts, so some sprinters use clothes to try to make themselves more aerodynamic.

Fish like this barracuda have evolved to have a streamlined body shape, reducing drag through the water.

More speed, more drag

Imagine walking along holding a large piece of card above your head. You would feel the resistance of the air as you walked. The air particles would push against you and the card, acting to slow you down.

If you started to run, then the drag would increase. You would feel a larger resistance. The faster something moves, the more drag it will experience. This is because when you move quickly, you collide with more air particles.

Guess what?

Drag causes heating. Spacecraft re-enter the Earth's atmosphere at very high speed. They collide with air particles so violently that their body glows red.
Teflon is a material that was discovered in 1949. It is used to coat the spacecraft body to protect it from the heat of re-entry. Similar materials are now used to coat non-stick frying pans.

Fuel consumption – *IT'S A DRAG*

Fed up of filling your car with petrol? The reason drivers need so much fuel is all to do with the forces opposing their motion, namely drag forces.

The engine burns fuel to produce a forward thrust at the wheels. Friction and air resistance are both forces that act against this forward thrust, creating drag. The amount of drag depends on the shape of the car and also on your speed. The faster you go, the more air particles you collide with and the more drag you experience.

Your fuel consumption depends on drag:

more speed → more drag → more thrust needed → more fuel consumed → more expense

The Model T did 20 miles per gallon. A modern family car is more streamlined and has a more efficient engine – it might do 50 miles per gallon. But you'll get more miles from each tankful of petrol if you slow down.

1 Copy and complete using words from the Language bank:

Air _____ and water resistance are _____ which oppose movement. They are also known as _____. Their effects can be reduced by _____ the shape. This gives a smaller surface for the _____ to collide with.

2 Why have designers tried to make cars more aerodynamic over the last 50 years?

3 The space shuttle's body becomes red-hot when it enters the Earth's atmosphere. Draw a diagram showing air particles to explain why this happens.

Language bank

aerodynamic
drag
forces
friction
particles
resistance
streamlining
speed
thrust
fuel consumption

A more detailed look at speed

How do parachutes work?

To understand the forces in parachuting, remember that air resistance depends on two things – shape and speed.

1 Two forces act when he jumps – air resistance and his weight. These forces are unbalanced. He is not moving very fast so air resistance is low at first. His weight is much larger so he gains speed, accelerating towards the Earth.

2 Now he is moving faster so the air resistance force is larger. Eventually it equals his weight. The forces are balanced so he is moving at a steady speed. This is the maximum speed that he can travel at in free fall, called the **terminal velocity**.

3 The parachute is designed to make the air resistance large. When it opens he is falling fast and the air resistance force is very big. The forces are unbalanced so the skydiver's speed changes. He keeps falling, but he falls more slowly.

4 He slows down till the force of air resistance again balances his weight. Again he falls at a constant speed, but more slowly this time. Slowing down is called **deceleration**.

5 Once landed the reaction force of the ground balances his weight.

1

2

3

4 air resistance

weight

5

reaction force ⬆ weight

The speed of fall depends on the balance of forces acting on the skydiver. His maximum speed in free fall is about 60 m/s or 130 miles per hour.

The skydiver is an example of what Newton was saying …

An object will remain stationary or at uniform (steady) speed, unless subjected to an external force.

The skydiver travels at a steady speed when:

the force of air resistance = the force of gravity (his weight)

Remember
On the Moon there is no air, so no air resistance. Falling objects would continue to accelerate as they are pulled down by gravity, even with a parachute … eek!

Drawing a graph of a journey

You may have seen graphs of distance against time for a journey. The slope of the line shows the speed.

Graphs of speed against time are similar, showing how the speed of a journey changes with time. They are sometimes called velocity–time graphs. The slope of the line shows the acceleration. A steep line means a fast acceleration.

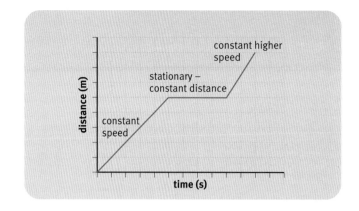

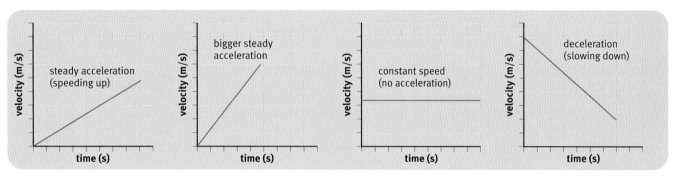

Look at this journey. Try to imagine what happened at each stage.

Velocity tells you the direction of movement as well as the speed.

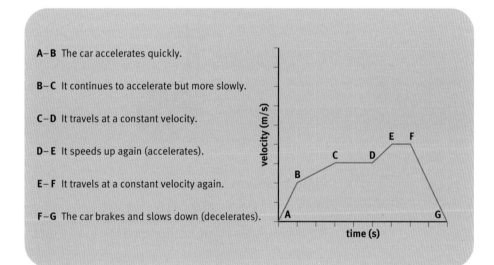

A–B The car accelerates quickly.

B–C It continues to accelerate but more slowly.

C–D It travels at a constant velocity.

D–E It speeds up again (accelerates).

E–F It travels at a constant velocity again.

F–G The car brakes and slows down (decelerates).

1 Copy and complete using words from the Language bank:

The force of air resistance depends on an object's shape and _____.
A parachutist falls at a constant speed when the upward force of _____ balances the downward force of his _____.

2 When does the parachutist stop accelerating? Explain why this happens.

3 A helicopter uses its rotor blades to create a 'down-draft' which pushes air downwards and lifts the helicopter. Why would a helicopter not work on the Moon?

Language bank

acceleration
air resistance
balanced forces
deceleration
drag
parachute
shape
speed
unbalanced forces
velocity
velocity–time graph
weight

Checkpoint

1 About speed

Match up the beginnings and endings to make complete sentences.

Beginnings

To find the average speed

An object will remain stationary or continue moving at a steady speed

An unbalanced force acting on a moving object

A change in speed

Endings

is called an acceleration.

we divide the total distance by the total time.

will cause a change in speed.

unless an unbalanced force acts on it.

2 What's the speed?

Copy and complete this table of running speeds.

Event	Time	Time (in seconds)	Average speed
100 m	9.92 s	9.92 s	10.1 m/s
200 m	19.54 s	19.54 s	
400 m	43.18 s	43.18 s	
800 m	1 min 46.11 s	106.11 s	
1500 m	3 min 32.00 s		

3 Slowing you down

a If you are moving, there are forces that slow you down. Unscramble the names of the forces below.

ari tresiancse

croifnit

gard

treaw crestisane

b Copy and complete the following sentences, choosing the correct words.

Air resistance is caused by **air particles / water particles / friction** bumping into a moving object.

Air resistance is **reduced / increased** if the shape is streamlined.

Air resistance is **reduced / increased** if you speed up.

4 Jump!

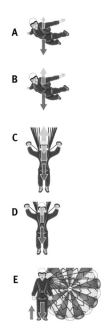

Choose the correct statement below for each label **A** to **E**.

Air resistance = weight: the speed is steady and lower.

Air resistance is lower than his weight: acceleration.

Weight = reaction force of the ground.

Air resistance is bigger than his weight: slowing down.

Air resistance = weight: the speed is steady and high.

Pressure and moments

Before starting this unit, you should already be familiar with these ideas from earlier work.

- A force is a push or a pull, and it has magnitude and direction. We show the direction with an arrow. How do we show the magnitude?
- To calculate area we multiply length by width. Find the area of a lawn 20 m wide and 35 m long. Don't forget the units!
- You move your body at joints, which is where bones meet. Muscles are attached to bones by tendons, and they pull on the bones. Can a muscle push a bone?

You will meet these key ideas as you work through this unit. Have a quick look now, and at the end of the unit read them through slowly.

- The effect of a force depends on not only the size of the force, but also on the area over which it acts. Force per unit area is **pressure**. We measure pressure in newtons per square metre.
- If we make the area small, e.g. a knife blade, the pressure is high. If we spread the force over a big area, as with skis, the pressure is low.
- A gas exerts pressure as its particles collide with the walls of its container, or with the surface of an object in the gas. **Atmospheric pressure** is smaller if you go high in the mountains than it is at sea level, because there is less weight of air above you.
- A liquid exerts pressure in the same way as a gas. Water pressure is greater the deeper you go.
- A lever is a simple machine. A lever turns around a pivot. For example, if you use a screwdriver to prise the lid off a tin you are using a lever.
- A force applied to a lever has a turning effect (we also call it a **moment**). The turning effect depends on the size of the force and its distance from the pivot. We measure moments in newton metres.
- When a lever is balanced, the moments are the same each side of the pivot. If the forces are not equal, their turning effects can be made to balance by moving the forces towards or away from the pivot.

(A)

What is pressure?

Under pressure

If this boy lay on just one nail, it would hurt him. But his weight is spread out on lots of nails and they do not pierce his body.

The reason for this is to do with **pressure**. Pressure is the effect of a force spread over an area. We calculate pressure by dividing the force (in newtons) by the area over which it acts (in square metres, m²).

pressure $= \dfrac{\text{force}}{\text{area}}$

The units we use to measure forces are newtons per square metre, N/m², which are sometimes called pascals, Pa.

Now you can see why the boy can lie on a bed of nails. For the same force (his weight), the pressure is smaller if the force is spread over a larger area.

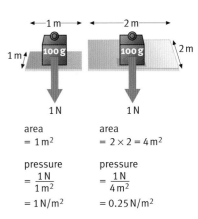

area
= 1 m²

area
= 2 × 2 = 4 m²

pressure
$= \dfrac{1\,\text{N}}{1\,\text{m}^2}$

pressure
$= \dfrac{1\,\text{N}}{4\,\text{m}^2}$

= 1 N/m²

= 0.25 N/m²

Remember that a mass of 100 g has a weight of 1 N (about an apple!).

Coastguard saves 7-year old from quicksand death

A YOUNG BOY failed to free himself after sinking up to his ankles in quicksand with the tide rapidly coming in. The coastguard team rose to the challenge of the hair-raising rescue operation. They placed a board around him to spread their weight. This stopped them from sinking into the notorious quicksand themselves. They pumped water into the sand below him to soften the quicksand so it would release him.

board

water pumped
into sand

The board made the pressure less concentrated.

No, it made the force less concentrated by spreading it out.

How much pressure do you exert?

Every year people like the boy in the article become stuck in quicksand or mudflats. In some types of quicksand you might sink if you exert a pressure greater than 30 000 N/m². So will *you* sink? What is the pressure that you exert on the ground under your feet?

You need to know your weight in newtons, and the area of your feet. Draw round a foot on graph paper and then count the squares.

Here is an example calculation:

mass = 40 kg
weight = 400 N
area of one foot = 128 cm^2 = 128 × 0.0001 m^2 = 0.0128 m^2
area of two feet = 2 × 0.0128 m^2 = 0.0256 m^2

$$\text{pressure} = \frac{\text{force}}{\text{area}} = \frac{400 \text{ N}}{0.0256 \text{ m}^2} = 31\ 250 \text{ N/m}^2$$

This person would sink!

Spreading out the force gives a lower pressure

In each of these photos the weight is spread out over a large area.

The surfboard spreads the surfer's weight, holding her on the water.

The snowshoes spread the weight, so he can walk on the snow.

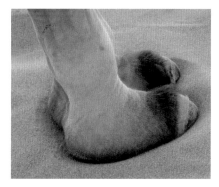

The camel has wide feet to stop it sinking into the sand.

Concentrating the force gives a higher pressure

Here the force is concentrated over a small area to increase the pressure.

The high pressure under the skates melts the ice. The film of liquid water formed underneath reduces the friction so you can glide.

The knife blade is very narrow. The force of your hand is concentrated and the pressure is high, so you can cut through the melon easily.

The drawing pin has a large head. The force of your push is concentrated at the point so it goes into the board.

1 Copy and complete using words from the Language bank:

The effect that a _____ has depends on the area to which it is applied. _____ is a force acting on a certain area.
Pressure = force ÷ _____ .

2 What is the pressure under a book weighing 6 N, area of 0.03 m^2?

3 Lofty repairs roofs. Explain why he puts a ladder over the roof tiles before standing on them.

Language bank

area
concentrated
force
newton per square metre
  (N/m²)
pascal
pressure
weight

○ **What is pneumatics?**

When you press the top of an aerosol can, gas escapes into the air. Inside the can is gas under pressure.

The same thing happens when you press the valve on a bicycle tyre. The air inside the tyre is at a higher pressure than the atmosphere. When you open the valve, air rushes out.

Pneumatics is all about using gases under pressure.

What is gas pressure?

You have seen how a force exerts pressure over an area. A gas can also exert pressure. Gases exert gas pressure by their particles hitting things.

Remember the particle theory? The particles in a gas are a long way apart. So gases can be compressed.

Look at the diagram. The first container is open to the air. Particles hit the container's sides, but they are hitting both the inside and outside at the same rate, so the pressure is equal.

The second container is sealed. The gas has been compressed. Particles hit the inside of the container more often than the outside. The pressure inside is higher.

The more you compress a gas, the more closely the particles are packed together and the more often they hit the container sides.

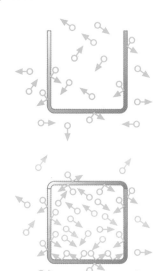

The particles of gas are moving rapidly and randomly. They collide with the sides of the container. These collisions cause pressure.

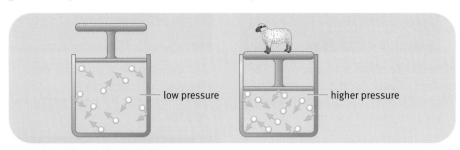

low pressure

higher pressure

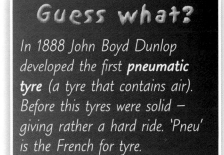

Guess what?

*In 1888 John Boyd Dunlop developed the first **pneumatic tyre** (a tyre that contains air). Before this tyres were solid – giving rather a hard ride. 'Pneu' is the French for tyre.*

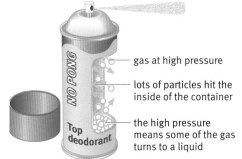

NO PONG

Top deodorant

gas at high pressure

lots of particles hit the inside of the container

the high pressure means some of the gas turns to a liquid

In an aerosol can or a bicycle tyre, the gas is compressed. The gas pressure inside is high. When you press the nozzle the particles rush out.

Using gases under pressure

In a steam engine, burning coal heats water and turns it into a gas (steam). Steam under pressure drives the train along.

Atmospheric pressure

The pressure of the air around us is called **atmospheric pressure**. At sea level this is 1 atmosphere or 100 kPa. You don't feel atmospheric pressure because there is high-pressure air in your lungs as well as outside. But if you go up a mountain the pressure is less, because there is less weight of air above you. You might feel breathless, because there is less oxygen available.

The butane gas in the cylinder is under so much pressure it has turned to liquid. When the burners are turned on the liquid escapes as a gas.

low pressure – less weight of air

high pressure – weight of air

Guess what?

Never put an aerosol can on a fire. The particles inside move even faster as it heats up, and the increasing pressure could make it explode.

Language bank

aerosol
atmospheric pressure
compressed
gas
gas pressure
liquid
particles
pneumatic tyre
pneumatics
pressure
tyre

1 Copy and complete using words from the Language bank:

   _____ is caused by gas particles hitting things. The _____ in a gas are far apart so gases can be _____ or squashed. _____ is all about using gases under pressure.

2 Explain why an inflated bicycle tyre feels harder than one that is not fully pumped up. Use the words 'particles' and 'pressure' in your answer.

3 Atmospheric pressure is lower at the top of a mountain. What might happen to an inflated balloon if you took it up a mountain?

○ **What is hydraulics?**

You can't squash a liquid

You know from the particle theory that the particles in a liquid are touching. So a liquid cannot be compressed like a gas can.

A liquid can be used to transmit pressure from one place to another. This is what **hydraulics** is all about.

The diagram below shows two syringes connected by a tube. The plunger in syringe A is small. The plunger in syringe B is big. Pushing down on A with a small force produces a larger force at B.

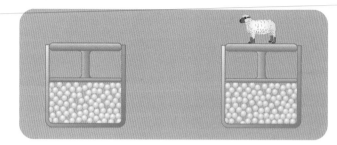

Using hydraulics

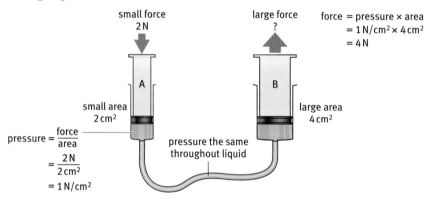

small force
2 N

large force
?

force = pressure × area
= 1 N/cm² × 4 cm²
= 4 N

pressure the same
throughout liquid

A

B

small area
2 cm²

large area
4 cm²

$$\text{pressure} = \frac{\text{force}}{\text{area}}$$

$$= \frac{2\,N}{2\,cm^2}$$

$$= 1\,N/cm^2$$

The pressure is the same throughout the liquid. It transmits the pressure from A to B.

Many hydraulic systems work like the syringes. Instead of syringes and plungers they have cylinders with pistons inside. The braking system on a car uses cylinders like this.

Guess what?

'Hydraulics' comes from **hydra** *meaning water and* **aulos** *meaning pipe. Water pipe?*

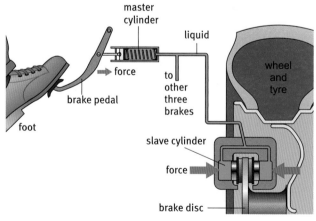

master cylinder

liquid

force

to other three brakes

wheel and tyre

brake pedal

foot

slave cylinder

force

brake disc

The driver presses the brake pedal, pushing in the master piston. The pressure is transmitted throughout the liquid in the pipes. This pressure exerts a large force on the slave pistons. The brakes slow the car down.

The robots use hydraulics to move.

Water pressure

You know that high in the mountains, the air pressure is lower than at sea level because there is a smaller weight of air pressing on you. In the same way, the deeper a diver goes, the bigger the weight of water above her and the higher the water pressure.

You can see this difference in water pressure in the photo below. The pressure is highest at the bottom of the bottle so here the water squirts the furthest.

The deeper you go, the higher the water pressure.

Guess what?

Divers breathe air at high pressure. Otherwise it wouldn't come out of the tanks in the high pressure of the deep water. This pressure causes nitrogen to dissolve in their blood. If they come up too fast the nitrogen makes bubbles in the blood, called 'the bends'. This is painful and can be fatal.

1 Copy and complete using words from the Language bank:

The _____ in a liquid are touching so you cannot _____ a liquid. A liquid _____ pressure from one place to another. This is what _____ is all about.

2 Divers' watches have to be specially sealed to keep the water out. They have a depth rating, e.g. 'waterproof to 200 m'.
   a What changes with water depth that could affect the watch?
   b Sketch what might happen if a diver goes too deep with the watch on.

3 A patient's drip bag is hung high on a stand. Liquid flows from the bag into their blood, to give them food or drugs. Explain why the bag is raised up, using the words 'water pressure' and 'weight'.

4 Find out what a manometer and an aneroid barometer are. Describe them and say what they are used for.

Language bank

atmospheric pressure
compress
depth
hydraulics
liquid
particles
pressure
transmits
water pressure

○ How do levers work?

A **lever** is a simple machine. A machine is something that helps us do work more easily . . . like taking the lid off a tin of toffees. A lever moves around a point called a **pivot**.

Your force on the lever is the **effort**. The weight of the thing you are moving is the **load**.

Using a screwdriver as a lever.

Levers in life

Look at these levers. The lever lets you do a job using a small effort force to move a larger load force.

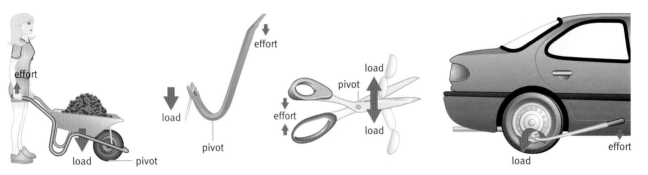

Making it easier

The longer the lever, the easier the job. This is because a longer lever gives a bigger turning effect.

turning effect = force × perpendicular distance from the pivot

We call the turning effect the **moment**. It is measured in units called newton metres, N m.

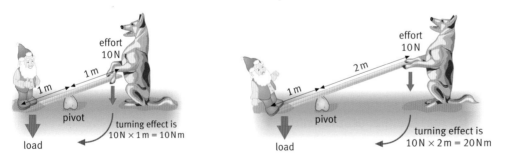

The dog does not have to push any harder, but the longer lever means he can lift the gnome.

A lever makes things easier by **amplifying** your force. It makes the turning effect of your force bigger. The longer the lever, the greater the turning effect.

Levers in the body

You move your body at joints. **Muscles** contract or shorten to pull on a bone. **Tendons** attach muscles to bones and transmit the force.

The girl with the longer arm has a longer lever. Can you work out where her pivot is?

Muscles work in pairs to move a bone. One muscle contracts to move the bone one way, and the other contracts to move it back. These muscle pairs are called **antagonistic muscles**.

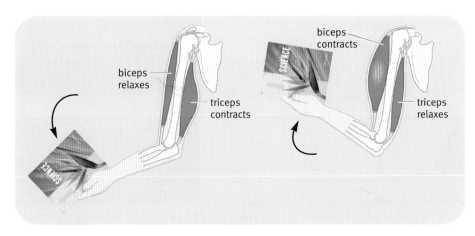

biceps relaxes

triceps contracts

biceps contracts

triceps relaxes

1 Copy and complete using words from the Language bank:

A lever is a simple _____ which uses a pivot. Your force is called the _____. The force you move is called the _____. The turning effect depends on the effort and its distance from the _____.

2 What do we mean by the turning effect of a force?

3 Sanjit opens a tin of paint using a 20 cm screwdriver. He pushes down with a force of 5 N. What is the turning effect of his force?

4 If you use a long handle on a wheel brace, it is easier to undo the nuts on a car wheel. Why is this?

Language bank

amplifying
antagonistic muscles
biceps
effort
force
lever
load
machine
moment
muscle
newton metre, N m
pivot
tendon
triceps
turning effect

| # In balance

Balancing the turning effect

The turning effect of the dog's force lifts the gnome. The turning effect is unbalanced so the seesaw turns.

We often want things to balance and not turn. When a crane lifts a heavy load, the turning effect of the load is huge. To stop the crane toppling over we need another turning effect on the other side to balance it.

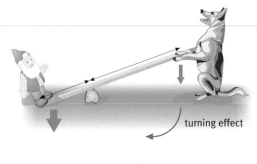
turning effect

This crane has a massive block of concrete called a **counterbalance**. Its turning effect balances the turning effect of the weight the crane is lifting.

For something to balance, the moments need to be equal on each side. We say that:

the anticlockwise moments = the clockwise moments

To change the turning effect on the seesaw, we can change either the force or the distance from the pivot (or both!).

counterbalance

load

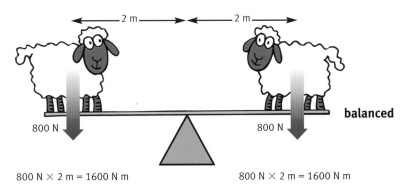

2 m   2 m

balanced

800 N   800 N

800 N × 2 m = 1600 N m   800 N × 2 m = 1600 N m

Anticlockwise moment equals clockwise moment and the seesaw is balanced.

> **Remember**
> The moment is just the turning effect, that is, force × distance from pivot.

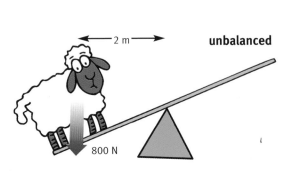

2 m   **unbalanced**

800 N

800 N × 2 m = 1600 N m   0

Anticlockwise moment is bigger than clockwise moment and the seesaw turns about the pivot.

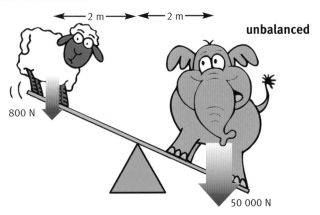

2 m   2 m

unbalanced

800 N

50 000 N

800 N × 2 m = 1600 N m   50 000 N × 2 m = 100 000 N m

Anticlockwise moment is smaller than clockwise moment and the seesaw turns about the pivot.

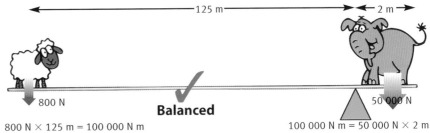

To change, the turning effect on the seesaw, we can change either the force or the distance from the pivot (or both!).

800 N
Balanced
50 000 N

800 N × 125 m = 100 000 N m   100 000 N m = 50 000 N × 2 m

A big distance increases the moment of your force.

Will it balance?

Will balance	Won't balance

Try it!

Don't try it . . .

Try it!

Guess what?

The best way to lift a heavy object is by bending your knees and keeping your back straight. This puts less strain on your back. You should also be close to the object to reduce its turning moment as you lift it. This is a good tip if you don't fancy back problems later in life.

Try to work out why each of these will or won't balance. First find the pivot. Think about moments and counterbalances.

1. Copy and complete using words from the Language bank:

   An object will topple over if there is an _____ turning effect. Another word for turning effect is _____. If the anticlockwise _____ equals the _____ moment, then the object will balance. If not then the object will turn about the _____.

2. Look at the picture of the crane. What do you think the operator does to the counterbalance when lifting loads of different weight?

3. Sarah pivots a metre rule at its centre. She hangs 15 N of slotted weights 20 cm from the pivot. Work out where she should hang 30 N of slotted weights on the other side to balance it.

4. Try and explain how the man on the chair is balancing. Sketch him and label the pivot. Draw arrows to show the turning moments on each side.

Language bank

anticlockwise
balance
clockwise
counterbalance
force
moment
pivot
turning effect
unbalanced

149

Checkpoint

1 About pressure

Decide whether the following statements are true or false. Write down the true ones. Correct the false ones before you write them down.

Pressure is the effect of a force spread out over a volume.

To calculate pressure we divide the speed by the area.

The units for force are newtons.

The units for area are circular metres.

The units for pressure are newtons per square metre.

Skis are designed to give a high pressure.

A knife blade is designed to give a high pressure.

2 Choose the answer

Copy and complete the following sentences, choosing the correct words.

You can compress a gas because its particles are **close together** / **far apart** / **touching**.

When a gas is compressed its particles hit the sides of the container more often, causing a higher **blood pressure** / **water pressure** / **gas pressure**.

Bicycle tyres contain gas at a **higher** / **lower** pressure than atmospheric pressure.

3 Complete the labels

Sketch the following diagram and complete the labels, working out the answer to each '?'.

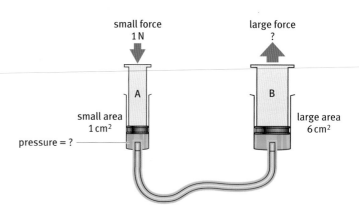

4 Levers

Match up the beginnings and endings to make complete sentences.

Beginnings

A lever moves around a point

Your force on the end of the lever is

The weight you are trying to move is

The turning effect or moment of a force is

If the moments are equal each side,

Endings

the load.

called the pivot.

a lever balances.

the effort.

the force multiplied by distance from pivot.

5 Will it balance?

For each seesaw **a** to **c**, work out the turning effect on each side.

Will it balance? If not, which way will it turn?

| | Left-hand side | | Right-hand side |
	Weight	Distance from pivot	Weight
a	400 N	1 m	1 m	400 N
b	300 N	2 m	1 m	300 N
c	400 N	2 m	1 m	800 N

Glossary

Words in italic have their own glossary entry.

A

acceleration A change in *speed* or velocity. Acceleration is measured in metres per second per second (m/s^2).

acid rain Rain which is more acidic than normal rainwater (i.e. rain with a pH lower than about pH 6).

activity series Another name for *reactivity series*.

addictive A *drug* is addictive if the person keeps wanting to take the drug.

aerodynamic Another name for *streamlined*.

aforestation Planting trees in an area that was previously *deforested*.

alcoholic A person who is dependent on alcohol.

alkali A soluble base.

ampere (A) Unit of *current*.

amplifying Making bigger. A *lever* amplifies a force by making its turning effect or *moment* bigger.

analogy comparing something we cannot see, such as how something works, with something simple that we know about, to make it easier to understand.

angina Pain in the chest, caused by insufficient oxygen reaching the heart muscle, because the arteries supplying the heart are blocked.

antagonistic pair A pair of *muscles* which act against each other, working together to produce movement at a *joint*.

artificial satellite A *satellite* made by people, such as a communications satellite.

artificial selection Another name for *selective breeding*.

asexual reproduction Producing offspring without the fusing of male and female sex cells. Asexual reproduction needs only one parent.

asthma A condition in which the airways become narrower, causing difficulty in breathing, often as a result of an allergic reaction.

atmospheric pressure *Pressure* which the atmosphere exerts on an object as the air particles collide with it.

average speed The average or mean *speed* over a whole journey. It is the total distance divided by the total time.

B

balanced diet A diet that provides the correct amounts of all seven types of *nutrient* needed for a healthy body.

ball-and-socket joint A *joint* such as the hip joint, where the bone can be moved in many directions.

base A substance (often a metal hydroxide or oxide) which will react with an acid to make a salt and water.

bioaccumulation The concentration of a substance building up in organisms as the substance passes through a *food chain*.

biological weathering Weathering of rocks caused by the action of plants or animals.

biomass The total mass of organic material in an organism, found by measuring the mass of the organism without the water.

breed A breed of plant or animal is a subgroup of the *species* which share *artificially selected characteristics* such as a similar appearance.

C

carcinogens Substances that cause cancer.

carnivore A meat-eating animal, for example a wolf or an osprey.

cartilage A smooth substance which covers the ends of bones in a *joint* and stops the bones rubbing on each other.

catalytic converter A device used in the exhaust systems of cars to reduce atmospheric pollution.

characteristics Typical features of an organism, which may or may not be passed on to the next generation.

chemical weathering Weathering of rocks caused by chemical reactions, e.g. with acids.

chest cavity Part of the body above the *diaphragm*, which contains the lungs.

chlorides Salts or compounds which contain chlorine, often formed from hydrochloric acid.

chlorophyll The green chemical in plants needed for *photosynthesis*.

chloroplasts Parts in a plant cell that contain *chlorophyll*. *Photosynthesis* takes place inside the chloroplasts.

ciliated epithelial cells Cells lining the airways, having tiny hair-like cilia which work to keep the *respiratory system* clean.

circulatory system All the organs of the body which work together to transport substances around the body, including the heart and blood vessels.

clay Very fine particles of weathered rock. A soil with lots of clay may be stiff and sticky, and does not drain well.

combustion Another name for burning or *oxidation*.

complete combustion A *combustion* reaction carried out in plenty of oxygen, so the fuel is fully oxidised.

conservation of mass A law which says that in a reaction, when reactants are converted to products, no mass is lost.

consumer An animal, that eats plants or other animals.

corrosion The reaction of a *metal* or other material with air and water.

counterbalance A *weight* used to balance another weight. For example, a crane uses a counterbalance to balance the *load* it is lifting and stop it toppling over.

crop yield How much of a crop plant is obtained per area of land.

current The rate of flow of electrical charge in a circuit.

D

deceleration A change in *speed* when an object is slowing down. A deceleration is a negative *acceleration*.

deficiency disease A disease caused by not eating enough of a certain *nutrient*.

deforestation Cutting down trees and not replacing them.

depressant A *drug* which makes people less alert, such as alcohol.

diamond A form of the element carbon which is very hard, used in jewellery and for the tips of drills.

diaphragm A sheet of muscle at the base of the *chest cavity* which is involved in *ventilation*.

digestive system All the organs of the body which work together to carry out digestion, including the stomach and the small intestine.

displace To take the place of another less reactive element in a compound.

displacement reaction A reaction in which a more reactive element takes the place of a less reactive one in a compound.

dissipated Energy is dissipated when it becomes *transferred* and *transformed* into types of energy that we cannot use. For example, a car engine dissipates energy as heat and sound.

drag Frictional forces which slow down a moving object such as a car.

drug A substance that affects the way the mind and/or body works.

E

efficiency How well a device *transforms* one type of energy to another.

effort The force you put in when you use a *lever* or other machine.

egg The female sex cell in animals or plants, which fuses with the male sex cell to produce offspring in *sexual reproduction*.

electrolysis Driving a chemical reaction using electricity, for example to extract a *metal* from its *ore*.

ellipse An oval shape. Many *orbits* are elliptical.

endothermic reaction A reaction which takes in energy.

energy transfer Energy moving from one place to another, e.g. from a battery to the components in a circuit.

energy transformation Energy being converted into another form of energy, e.g. an electric heater transforms electrical energy into heat energy (and some light energy).

environmental variation Variation that comes from characteristics that are not passed on from parents, but depend on the environment or conditions in which the organism lives.

exothermic reaction A reaction which releases energy.

F

feeding relationships The way organisms are linked in a *food chain* or *food web*.

fertilisers Chemicals which are added to soil to provide the correct *minerals* for plants to grow well.

food chain A series of organisms that are dependent on one another for food. Each organism is eaten by the next in the chain.

food web All the *food chains* in a community joined together in a network.

fractional distillation Separating a mixture such as crude oil into fractions with different boiling points.

fuel A chemical which is burnt to release energy.

G

generator A device that *transforms* kinetic energy into electrical energy.

genes Found in chromosomes in the nucleus, genes contain information about the *characteristics* of an organism.

geocentric model A model of the Solar System in which the Earth is placed at the centre.

geostationary orbit An *orbit* around the Earth in which a satellite remains in one position over the Earth's surface, orbiting at the same speed as the Earth rotates.

global warming Theory that the increased *greenhouse effect* is causing climate change on the Earth.

glucose A simple sugar which is used in respiration and produced by plants in *photosynthesis*.

graphite A form of the element carbon, a *non-metal* that is unusual because it conducts electricity. Graphite has a layered structure and is used in pencils.

gravitational force The size of the force of *gravity* between two masses.

gravity A force of attraction that acts between objects that have *mass*. Earth's gravity keeps everything on Earth from floating out into space.

greenhouse effect The action of *greenhouse gases* in the atmosphere, which act like the glass in a greenhouse, reflecting infrared radiation back to Earth and so making the Earth warmer than it would otherwise be.

greenhouse gases Gases that contribute to the *greenhouse effect*, including carbon dioxide, water vapour and methane.

H

hallucinations Seeing something that is not really there, but it seems real.

heart attack Part of the heart dies when insufficient oxygen reaches it, because the arteries supplying the heart are blocked.

heliocentric model A model of the Solar System in which the Sun is placed at the centre.

herbicide A chemical used to control or kill plants (another name for a *weedkiller*).

herbivore A plant-eating animal, for example a sheep or a cow.

high blood pressure The blood in the blood vessels being at a higher pressure than normal, which can cause health problems.

hinge joint A *joint* such as the elbow joint, where the bone can be moved in one direction only (like a hinge).

humus The organic material in soil, made from dead and decayed plants and animals.

hydraulics The study and use of liquids to transmit forces, using cylinders and pistons.

hydrocarbons Compounds which contain hydrogen and carbon only.

I

illegal drugs Harmful *drugs* taken for recreation, which are against the law to take or possess.

incomplete combustion A *combustion* reaction carried out without enough oxygen, so the fuel is not fully oxidised. This can be dangerous as the reaction can produce carbon monoxide.

indicator organism An organism which can tell scientists about the level of pollution in a certain area.

inherited variation Variation that comes from *characteristics* that are passed from parents to their offspring, in the *genes*.

J

joint A place where bones meet and may be moved against each other.

L

leaf The organ of a plant which is well designed to carry out *photosynthesis*.

lever A simple machine used to do work more easily, such as removing a tin lid.

ligament Strong fibres which connect bone to bone at a *joint*.

liver The organ which removes toxic materials such as alcohol from the blood.

load The force you move when you use a *lever*.

M

mass A measure of the quantity of matter that something contains. Mass is measured in kilograms.

medicinal drugs Drugs such as antibiotics or anti-inflammatories, that are taken to treat illnesses.

melanin The pigment in human skin which causes it to go brown when exposed to sunlight, and protects us from ultraviolet light.

mercury The only *metal* which is a liquid at room temperature.

metals Elements found on the left-hand side of the periodic table, which are shiny, hard and feel cold to the touch.

minerals/mineral salts Compounds which contain important elements needed for health in animals or plants, e.g. iron sulphate is a mineral salt that contains the element iron.

model A way of picturing something we cannot see, such as a complex theory, to make it easier to understand.

moment The turning effect of a force, found by multiplying the force by the perpendicular distance from the pivot.

monomers The repeating units in a *polymer* molecule.

muscle An organ that contracts to move a bone at a *joint*.

N

native A native *metal* is found as the element in nature, rather than combined in an *ore*.

natural selection Certain inherited *characteristics* are well suited to the natural conditions, and these are passed on to the next generation (survival of the fittest).

nitrates Salts or compounds which contain nitrogen and oxygen, often formed from nitric acid.

non-metals Elements found on the right-hand side of the periodic table, which do not have the properties of *metals*.

non-selective herbicide A *herbicide* that kills a wide range of plants, or all the plants it reaches.

nutrients Substances in foods that supply the chemicals needed for a healthy body.

O

omnivores Animals that eat both plants and other animals.

orbit The path a *satellite* (such as a planet) takes around the object it is orbiting (such as the Sun).

ore A compound containing a *metal*, from which the metal has to be extracted.

oxidation A reaction in which something reacts with oxygen.

oxide A compound that contains an element chemically joined to oxygen.

ozone layer A layer of ozone gas high in the atmosphere, which blocks much of the harmful ultraviolet light in sunlight and protects us from its effects.

P

palisade mesophyll A tissue in a *leaf* made up of column-shaped cells with lots of *chloroplasts*, where most *photosynthesis* happens.

passive smoking Breathing in the second-hand smoke produced by others.

pest An organism that reduces the *yield* of a crop, competing with humans for food.

pesticide A chemical used to control or kill *pests*.

photosynthesis The process by which plants change carbon dioxide and water into glucose, using sunlight and chlorophyll.

physical weathering Weathering of rocks caused by repeated heating and cooling, or ice in cracks melting and freezing.

pivot The point around which a *lever* turns.

pneumatic tyre A tyre filled with compressed air to cushion the ride.

pneumatics The study and use of gases under *pressure*.

pollen The male sex cell in plants, which fuses with the female sex cell to produce offspring in *sexual reproduction*.

polymers Materials containing long molecules made from many repeating units called *monomers*.

potential difference The amount of electrical energy carried by each bit of charge in an electrical circuit. Also called *voltage*, it is measured by a voltmeter.

pressure The effect of a force spread out over an area. Pressure is force divided by area, measured in newtons per square metre (N/m^2).

primary consumers The first level of consumers in a *food chain*, which eat plants.

producer A green plant, which produces it own food and is found at the start of a *food chain* or *food web*.

pyramid of biomass A diagram showing the total *biomass* of living material in each level of a *food chain*.

R

reaction time The time it takes for you to react to something.

reactivity How reactive something is, i.e. how quickly or readily it reacts.

reactivity series A list of elements (usually *metals*) in order, from most reactive to least reactive.

recovery rate The time it takes for your heart rate (pulse rate) to return to normal after exercise.

recreational drugs *Drugs* people take because they enjoy them, such as caffeine.

respiration The process by which living things release energy from their food, using oxygen.

respiratory system All the organs of the body which work together to carry out gas exchange, including the lungs and airways.

rib cage The bony cage formed by the ribs that surrounds the *chest cavity*.

root The organ of a plant which anchors it in the soil, and absorbs water and *minerals*.

root hairs Tiny hairs on the surface of cells on the *roots* of plants, which help them absorb water and *minerals*.

rust Flaky red iron oxide, formed by the *corrosion* of iron.

S

sand Tiny particles of silicon dioxide, often formed from weathered quartz. A soil with lots of sand drains well.

satellite An object (natural or artificial) that *orbits* another object.

scrubber A device used in the chimneys of factories and power stations to reduce the emission of polluting gases.

secondary consumers The second level of *consumers* in a *food chain*, which eat *primary consumers*.

selective breeding Choosing parent organisms with certain *characteristics* and mating them to try and produce offspring that have these characteristics.

selective herbicide A *herbicide* that kills a narrow range of plants.

selective pollination Taking *pollen* from a certain flower and placing it on the stigma of another, to carry out *selective breeding* in plants.

sexual reproduction Producing offspring by the fusing of male and female sex cells.

side effects Effects of using a *drug* other than the effect it is designed to give.

smelting Extracting a *metal* from its *ore* by heating the ore with carbon.

specialised cells Cells that are well suited to a certain job.

species A group of similar organisms that can breed together to produce fertile offspring.

speed A measure of how fast something is moving. Speed is the distance moved divided by the time taken, often measured in metres per second (m/s).

sperm The male sex cell in animals, which fuses with the female sex cell to produce offspring in *sexual reproduction*.

spongy mesophyll A tissue in a *leaf* made up of cells with spaces between them, where gases can move between the cells.

stimulant A *drug* which makes people feel more energetic or alert, such as amphetamines.

stoma (plural stomata) A hole in the underside of a *leaf*, through which gases can move in and out of the leaf.

streamlined A shape which reduces *drag*.

stroke A *thrombosis* which interrupts the blood flow to the brain, causing brain damage.

sulphates Salts or compounds which contain sulphur and oxygen, often formed from sulphuric acid.

sustainable development Developing a country or area without destroying the environment.

synovial fluid A liquid found in many *joints* which lubricates the joint.

synthesise To make or produce a substance.

synthetic Made by people, not found naturally in the environment.

T

tarnished Describes a *metal* that has gone dull due to reaction with the air and water.

tendon Strong fibres which connect muscle to bone at a *joint*.

terminal velocity The fastest possible speed when falling to Earth.

thermit reaction A *displacement reaction* between aluminium and iron oxide to produce molten iron, which is used to weld railway lines.

thrombosis A blood clot, usually in a blood vessel.

toxins Chemicals which are poisonous.

transfer *See* energy transfer.

transformation *See* energy transformation.

turbine A large blade often used to turn a *generator*.

U

unit of alcohol Unit used to measure the amount of alcohol people drink each week. One small glass of wine or half a pint of beer contains about one unit of alcohol.

unreactive Describes a substance that does not readily react.

V

vein In a plant, a tube-like structure which carries water and dissolved substances around the plant.

ventilation Another name for breathing, or moving air into and out of the lungs.

volt (V) Unit of *voltage*.

voltage Another name for *potential difference*.

W

weed A plant that competes with a crop plant – a plant in the wrong place.

weedkiller A chemical used to control or kill *weeds*.

weight The force with which a *mass* is pulled towards the centre of the Earth as a result of *gravity*.

withdrawal symptoms Bad effects that people experience if they stop taking an *addictive drug*.

Index